Foundations of a Sustainable Market Economy

Margreet Boersma-de Jong · Gjalt de Jong

Foundations of a Sustainable Market Economy

Guiding Principles for Change

Margreet Boersma-de Jong
Norg, The Netherlands

Gjalt de Jong
Norg, The Netherlands

ISBN 978-3-031-28185-3   ISBN 978-3-031-28186-0 (eBook)
https://doi.org/10.1007/978-3-031-28186-0

© The Editor(s) (if applicable) and The Author(s), under exclusive license to Springer Nature Switzerland AG 2023

This work is subject to copyright. All rights are solely and exclusively licensed by the Publisher, whether the whole or part of the material is concerned, specifically the rights of reprinting, reuse of illustrations, recitation, broadcasting, reproduction on microfilms or in any other physical way, and transmission or information storage and retrieval, electronic adaptation, computer software, or by similar or dissimilar methodology now known or hereafter developed.

The use of general descriptive names, registered names, trademarks, service marks, etc. in this publication does not imply, even in the absence of a specific statement, that such names are exempt from the relevant protective laws and regulations and therefore free for general use.

The publisher, the authors, and the editors are safe to assume that the advice and information in this book are believed to be true and accurate at the date of publication. Neither the publisher nor the authors or the editors give a warranty, expressed or implied, with respect to the material contained herein or for any errors or omissions that may have been made. The publisher remains neutral with regard to jurisdictional claims in published maps and institutional affiliations.

Cover credit: Tom Penpark

This Palgrave Macmillan imprint is published by the registered company Springer Nature Switzerland AG
The registered company address is: Gewerbestrasse 11, 6330 Cham, Switzerland

Preface

This textbook grew out of a genuine sense of wonder about our struggle to create a sustainable world and economy. Why do consumers persist in purchasing inexpensive meat despite their awareness of the inherent animal suffering involved? Why have we allowed climate change to come this far? The term 'market forces' is often used to answer these questions, but what are market forces? We read about Adam Smith, globalisation and other characteristics associated with markets. When we listed these characteristics and shared them in lectures with students and colleagues, we realized that we did not need to add anything to the table of contents of this book. It became clear how the market and an unsustainable world is connected, and what is needed to change it into a 'sustainable market economy'.

Sustainable market economy is introduced here as a counterpart to 'free market economy'. Unlimited individual freedom cannot exist in a world with limited resources in which everyone is equal. We cannot build a sustainable economy on freedom that leads to exploitation of fellow human beings and depletion of the planet.

In this book you will often come across the words 'we' and 'us'. These words were chosen intentionally to bring the subjects as close to your heart as possible. It is not *the* government; it is *our* government. It is we who elect the representatives of the people. It is not *the* economy; the economy belongs to all of us, and our behaviour determines whether we allow a free market economy to exist or whether we transition to our sustainable market economy.

Norg, The Netherlands   Margreet Boersma-de Jong
February 2023   Gjalt de Jong

Introduction

Greenhouse gas concentrations are at their highest levels in 2 million years, and emissions continue to rise. As a result, the Earth is now about 1.1 °C warmer than it was in the late 1800s. The last decade (2011–2020) was the warmest on record.[1] Although climate change is on the global agenda, we are not succeeding in creating sustainable change. How can this be done? And what role does the economy play in this?

This book teaches you about the relationship between economics and sustainability. Market forces have long been central to economics. This means that supply and demand are matched by means of price (you will learn more about this in Chapter 1). You often come across discussions in the news about various related topics, such as the benefits and drawbacks of market forces in healthcare or the ongoing debate about the value of free international trade and economic growth.

Markets have facilitated the development of numerous beneficial advancements and outcomes, but they also play an important role in the sustainability problems we now face. Climate change, child labour and oil spills are just a few of the many problems related to our economic activities.

Many consumers now understand that we need to change our economy. Many companies are trying to produce more sustainably, but progress is slow. What forces impede a quick transition towards a more sustainable economy?

---

[1] What Is Climate Change? Accessed September 30, 2022, from https://www.un.org/en/climatechange/what-is-climate-change

This question is central to the first part of this book. It turns out that our market, which in many countries is left to operate as freely as possible, is not as free as it seems. For instance, reusing raw materials is better for the earth, but is often more expensive than extracting more raw materials. Consequently, customers are required to pay a premium for products incorporating recycled materials. Organic meat and sustainable clothing are also more expensive than what we are used to. To combat climate change, we must switch en masse to clean energy. Hydrogen is promising, but supply is still low because suppliers do not yet have any customers. Customers are only willing to buy hydrogen when they are sure that this is a sustainable technology. All these examples indicate that the market falls short in fostering positive global change.

Hence, our market forces driven economy is also a compelling system. Companies *must* make certain choices to survive. You will learn more about this in Chapter 1. Next, Chapter 2 shows the negative effects of these choices, such as poor animal welfare, burn-out and social differences between rich and poor. In Chapter 3, you learn which guiding principles underlie our economic system. The market originated from human needs, yet it is also influenced by our perceptions of the market and the agreements we establish. In this chapter you learn about freedom, economic growth and humans as means of production. This knowledge is essential to understand why a more sustainable world and economy is struggling to get off the ground. A sustainable world is one in which the needs of the current generation are met without compromising the ability of future generations to meet their needs (this definition comes from the Brundtland report, which you can read more about in this book). Organisations often take well-intentioned sustainable actions, only to discover how strong the coercive hand of the market is. Sustainable guiding principles are needed to support businesses to make the transition. A sustainable market means accepting the limits on freedom set by the earth and other human beings. From that guiding principle, in combination with seven other principles discussed in Chapter 4, Chapter 5 shows the business models that can lead to a sustainable world.

Tip

The prevailing business paradigm places a strong emphasis on financial gain, with many individuals associating wealth with a beautiful and fulfilling life. While a basic income is undoubtedly important, it is crucial to explore what truly constitutes a fulfilling life. In the book "More than Money" by Mark Albion, there is a captivating parable about a fisherman and a businessman that delves into this concept. You can also find an accompanying animation of the parable on YouTube, adding a visual dimension to the story.

Contents

1	**The Functioning of the Free Market**	1
1.1	What Are Market Forces?	2
	Consumers prioritize three essential criteria: Innovation, Quality and Price	5
	Shareholders strive for high returns on their investments	9
1.2	Effects of the Compelling Hand on Business Choices	13
	Investing in Product Leadership	14
	Excellent Business Processes Through Cost Savings	16
1.3	Summary	24
	Literature	27
2	**The Tragedy of the Compelling Hand**	29
2.1	Damage Caused by Product Leadership	29
	Overexploitation	32
	Damage to Humans, Animals and the Earth	34
	Waste	39
	Damage arising from a Focus on High ROI and Patent Rights	41
2.2	Damage arising from Cost Savings	42
	Damage Caused by Efficient Use of Resources	43
	Damage Caused by Managing Employee Productivity	46
	Damage Caused by Keeping Labour Costs Low	53
2.3	Summary	60
	Literature	64

xii   Contents

3	**Guiding Principles in Our Current Economic Model**	67
	3.1 Acting in Freedom for More Possession	69
	Freedom to Serve One's Own Interests	69
	Labour Specialization for Higher Productivity	70
	A Larger Market Through Free Trade	71
	Right of Ownership Aimed at Private Property	72
	Monetary Value Drives Our Activities	73
	3.2 The Employee Is a Means of Production	76
	Employer–Employee: Who Pays, decides	77
	People Need to Be Motivated to Be Productive	81
	3.3 Economic Growth and *Trickle Down*	84
	3.4 The Role of the Government in the Market Economy	91
	3.5 Summary	94
	Literature	97
4	**Guiding Principles for a Sustainable Economy**	99
	4.1 Acting Responsibly for Greater Well-Being	99
	Responsible Self-Interest Through Ethical Awareness	101
	Interdisciplinary Work for Sustainable Solutions	112
	Local Trade and International Sustainable Chains	113
	Ownership and the Sharing Economy	115
	Multiple Values Guide Our Activity	118
	4.2 People Are Equal Partners in Production	120
	Equality Between Production Partners: Who Contributes, Decides	122
	People Are Motivated to Be of Value	123
	4.3 Multiple Growth for All	124
	4.4 The Role of Government in a Sustainable Market	128
	4.5 Summary	132
	Literature	137
5	**The Happiness of a Sustainable Market**	139
	5.1 How the Sustainable Market Works	139
	5.2 Translating Guiding Principles into a Sustainable Mission and Vision	143
	5.3 Sustainable Product Innovation and Improvement	146
	From Linear to Circular	146
	A Sustainable Return on Investment	153
	The Power of Marketing	154
	5.4 Cost Savings: Making Production Processes More Sustainable	154

		What Are Costs in the Sustainable Economy?	155
		Multiple Cost Reduction: Sustainable Use of Resources	161
		Equivalence: Joint Decision-Making	163
		Equality: Everyone Contributes	169
	5.5	Summary	170
	Literature		173

Afterword: Taking Action! The Role of Individuals, Businesses and Government   175

Literature   177

Index   181

About the Authors

Margreet Boersma-de Jong studied business administration at the University of Groningen in The Netherlands, where she obtained her doctorate in 1999 with a thesis on the development of trust between partners in international joint ventures. After 4 years of organizational consulting at PricewaterhouseCoopers, she switched to health care. In the 9 years that followed she advised and supervised teams on strategic financial matters. In 2012 she made the switch to the Hanze University of applied sciences Groningen, where she resumed her research as a Professor of Sustainable Finance and Entrepreneurship.

Gjalt de Jong is Professor of Sustainable Entrepreneurship in a Circular Economy (University of Groningen, Faculty Campus Fryslân). Professor de Jong builds on more than 30 years of experience in academia, business and management. He holds a Master of Science in Economics and a Ph.D. in Business Administration from the University of Groningen. He is the chair and head of the department Sustainable Entrepreneurship in a Circular Economy, founder-former director of the Master of Science program Sustainable Entrepreneurship, founder-director of the Centre for Sustainable Entrepreneurship, founder of the Sustainable Start-up Academy, and chair of the board of the Faculty's research institute. He currently also serves as the strategic advisor for the President of the University of Groningen concerning systemic transitions in agriculture. Prior to his current position he served as an associate professor at the University of Groningen (Faculty of

Economics and Business) and as a senior management consultant for KPMG and PricewaterhouseCoopers.

Foundations of a Sustainable Market Economy arose from our genuine wonder about the struggle our society has to make the world more sustainable. Why do consumers buy cheap meat, when they know that animals have to suffer for it? Why do we let climate change get this far? A puzzle that has occupied us both professionally and privately during our working life. With this book, we want to show students and other interested parties the true and harmful workings of the market and we present the new DNA of a sustainable market economy.

List of Figures

Image 1.1	Melting icebergs due to climate change (*Source* Pixabay, Markus Kammermann)	1
Image 1.2	In many markets the products do not differ much in quality and price. So how do you sell more than your competitor? Entire studies have examined this question, resulting in the field of marketing (*Source* Pixabay, Martin Winkler)	3
Fig. 1.1	The three strategies of Treacy and Wiersema (1995)	6
Image 1.3	Consumer society (*Source* Pixabay, StockSnap)	7
Image 1.4	An old model of Nokia (*Source* Kostkarubika005)	8
Image 1.5	The stock market is an attractive but risky way to increase your capital (*Source* Pix1861)	11
Image 1.6	One of the oldest shares of the VOC, dated 1606, in the name of the grocer Theunis Jansz for six hundred guilders (*Source* Pixabay Amsterdam City Archives)	12
Fig. 1.2	Customer and shareholder influence corporate choices	13
Image 1.7	Rusk with a notch. The notch on the side of the rusk, where exactly your finger fits in, was also granted patent protection (*Source* Own image)	15
Image 1.8	Mining for raw materials such as copper and lythium leaves deep scars in nature (*Source* Pixabay, Shibang)	18

List of Figures

Image 1.9	A Boeing 747-400 holds 408 passengers (*Source* Pixabay, PatrickE)	20
Fig. 2.1	Linear use of resources	31
Fig. 2.2	Overview of the date when as many of the Earth's resources were used as the Earth can give (*Source* Global Footprint Network[1])	33
Fig. 2.3	The Planetary Boundaries overview, designed by Azote for Stockholm Resilience Centre (*Source* Stockholm Resilience Centre, CC BY 4.0)	34
Image 2.1	This is what Pripjat (a place near Chernobyl) looks like: abandoned and overgrown (*Source* Pixabay, Wendelin_Jacober)	36
Image 2.2	Deforestation (*Source* Pixabay, Camera-man)	38
Image 2.3	A calf (*Source* Pixabay, RyanMcGuire)	44
Image 2.4	Insects need nature without pesticides (*Source* Own image)	45
Fig. 2.4	The two dimensions for job happiness from Singh and Aggarwal	49
Image 2.5	We do not have to shoulder the burdens of the world alone (*Source* Pixabay, StockSnap)	52
Image 2.6	The Jewish star, obligatory to wear on clothing by Jews in the World War II (*Source* Pixabay, JordanHoliday)	54
Fig. 2.5	Income inequality around the world (*Source* Own graph based on figures of the OECD[25])	55
Image 2.7	Homeless man fighting societal prejudices (*Source* Pixabay, IslandWorks)	57
Fig. 3.1	The economic system	68
Image 3.1	A poster of a pin factory in the time of Adam Smith. The image originally appears in Diderot's Encyclopedia (1751–1772)	70
Image 3.2	Hierarchy in organizations (*Source* Pixabay, geralt)	78
Image 3.3	Controlled motivation means that an organization devises external incentives that make people want to achieve a particular goal (*Source* Pixabay, mohamed_hassan)	83
Fig. 3.2	The eight guiding principles of a market economy	84

List of Figures

Fig. 3.3	A graph by Hans Rosling, relationship between income and life expectancy (*Source* Based on a free chart from www.gapminder.org/teach)	86
Fig. 3.4	In an economy, growth and recession alternate every few years	90
Fig. 3.5	The spectrum of the market on which government operates	91
Fig. 4.1	Smith argues that self-interest maximization is constrained by laws and religion, informal norms and values and empathy	102
Fig. 4.2	People, Planet, Profit: sustainable companies produce goods and services that generate profit for themselves, for the earth and for their fellow human beings	108
Fig. 4.3	Kate Raworth's donut model (*Source* Wikimedia Commons—DoughnutEconomics)	109
Image 4.1	The United Nations' seventeen Sustainable Development Goals (*Source* United Nations)	110
Image 4.2	Our products are transported all over the world before they arrive in your home (*Source* Pixabay, plindena)	114
Fig. 4.4	The difference between an unfair price and a fair price(*Source* Picture created by myself, courtesy of the four students)	119
Image 4.3	Reducing inequality is one the SDG's (*Source* United Nations)	123
Fig. 4.5	Eight guiding principles of a sustainable market economy	126
Fig. 4.6	The OECD dashboard to map well-being in a given year and future well-being (*Source* OECD)	127
Image 4.4	A fair price for milk can increase biodiversity (*Source* Pixabay, CallyL)	129
Image 5.1	Small-scale farming has declined sharply (*Source* Pixabay, analogicus)	142
Fig. 5.1	Companies can operationalize sustainability in three ways	143
Image 5.2	The impact of Tony's Chocolonely (*Source* Tony's Chocolonely)	144
Fig. 5.2	Sinek's golden circle	146
Image 5.3	Methods for circular living and business (*Source* PlanBureau voor de Leefomgeving (PBL))	148

Image 5.4	Windmills are a highly effective and environmentally friendly means of generating clean energy (*Source* Pixabay, jwvein)	150
Fig. 5.3	To speak of a sustainable product, three aspects are important	152
Image 5.5	Donating trees to offset your carbon footprint (*Source* own image)	156
Fig. 5.4	Life Cycle Analysis	159
Fig. 5.5	Sustainable cost savings	160
Fig. 5.6	The basic design of a sociocratic organization	164

List of Tables

Table 1.1	Firms' competitive strategies	6
Table 3.1	A trade-off between costs and benefits	75
Table 5.1	The different strategies with explanations	149

1

The Functioning of the Free Market

Climate change is one of the biggest challenges of our time. Temperatures are breaking records every year. We see more forest fires and floods than ever before. While there is a broad consensus on the influence of our actions on climate change, changes in behaviour and policy are slow in coming, and perhaps too late. The free market plays an important role in this problem: 'According to economists like Nicholas Stern, the climate crisis is a result of multiple market failures'[1] (Image 1.1).

Image 1.1 Melting icebergs due to climate change (*Source* Pixabay, Markus Kammermann)

Every day, enormous quantities of products are produced, transported and consumed. Production and consumption, supply and demand, are two essential parts of the concept of the market. In the market, the interplay of supply and demand is equilibrated through the mechanism of price. These mechanisms are commonly referred to as market forces or free market forces. These forces provide an important means of survival and make our lives more comfortable. However, market forces also have disadvantages. You will learn more about the pros and cons of the market in this book.

To properly understand these advantages and disadvantages, we must first understand what market forces are exactly. We first explore these forces in the context of a free market, one in which the government does not apply restrictive measures. Then we look at which strategies companies use to survive in a free market. This knowledge will provide the groundwork needed for Chapter 2, where we examine how these strategies are linked to the theme of sustainability.

1.1 What Are Market Forces?

To understand market forces, you can start with a simple idea: your own market in your town or village. Local markets are the most tangible, concrete manifestations of the concept of a market. Merchants enter the square with their stalls selling vegetables, fruit, cheese, nuts, fish and more. While wandering through the marketplace, consumers have the opportunity to compare different products and ultimately select the most suitable option, potentially influenced by the advertising strategies employed by the merchants. Preferences for certain stalls often emerge, based on previous purchases that met expectations (Image 1.2).

1 The Functioning of the Free Market

Image 1.2 In many markets the products do not differ much in quality and price. So how do you sell more than your competitor? Entire studies have examined this question, resulting in the field of marketing (*Source* Pixabay, Martin Winkler)

Markets have been around for centuries. To survive, people traded their livestock, food and other products so that together they had more than each on their own. Markets were and are used both as a means to survive and as a way to make life more pleasant. What happens in each market square also happens regionally, nationally and globally. Entrepreneurs offer goods at a certain price and quality, and consumers look for the best bargains or the goods with the best value for money. The marketing of the products plays an important role in this. In order to make their products stand out, entrepreneurs use clever seduction tactics.

With the rise of the Internet, the marketplace has expanded enormously. Until the late twentieth century, consumers had to physically go out and buy products. Nowadays, online stores enable us to make a few clicks on a tablet and have our order delivered the next day. The Internet has therefore greatly reduced the cost of searching for the right products; instead of spending hours in shops looking for that one unique shirt, we now scroll through the images on our screens.

We buy and sell things every day. What is special about the market is that it is not something invented by someone. It came about because both sellers and buyers had an interest in it. Bakers don't just bake bread for themselves but for the whole village. They only do that because they get something in return:

money to buy other products. As a consumer, you buy products because it is in your own interest. You buy things because you need them to survive or because they make your life better. Thus, the market was and is a means of serving this mutual interest.

Adam Smith (1732–1790) studied markets and coined the term *the invisible hand* of the market: the pursuit of one's own interests leads to an increase in prosperity for everyone. Indeed, if individuals collectively produce and exchange a surplus of their respective products through the market, it ultimately leads to an increase in the overall availability of goods for everyone. You can read more about his ideas in Chapters 3 and 4.

A market is called free if supply and demand are in balance without government intervention. The price is the regulating mechanism in that market: if there is much demand and little supply, producers can ask for higher prices. For example, after the coronavirus shutdowns, many people wanted to rent a car for holidays in 2022. Additionally, there was a shortage of rental cars in some areas. This led to a week of car rental costing around 1600 euros compared to around 800 euros before the pandemic. The reverse also occurs: if there is much supply but little demand, providers tend to lower the price to still receive some income.

A completely free market is theoretical. In reality, governments often play a regulatory role. Nevertheless, the free market is often discussed, particularly in contrast with economies where governments play a much larger role in matching supply and demand. Markets and free markets are therefore used interchangeably in this book.

From Oxford Learners Dictionary
Free market
noun
/ˌfriː ˈmɑːkɪt/
/ˌfriː ˈmɑːrkɪt/
an economic system in which the price of goods and services is affected by supply and demand rather than controlled by a government.

The market contains various players. This book focusses on three players: *customers*, who make choices about what they buy, and *shareholders* (or financiers) who decide what they want to invest their money in. Together they influence the choices *companies* make. The next sections will elaborate on each party separately.

Consumers prioritize three essential criteria: Innovation, Quality and Price

Suppose you live in the distant past, where each villager offers their own product. The exchange is simple: two loaves of bread for one pot of jam, one bearskin for building a sod hut. As long as there is only one producer of each product, consumers will have to buy what is available. Such a monopoly is advantageous for the producer: if you are the only one who makes a product, you can demand a high price. For consumers, a monopoly is therefore disadvantageous. Today most products have many suppliers. So how do they make sure that people continue to buy their product? Producers apply various types of strategies:

- They lower the price. This often means cutting costs if they want to end the year with the same profit margin.
- They invest in the quality of the product, distinguishing it from their competitors'.
- They innovate, again giving them a unique product.
- They build optimal customer relationships by investing in customer attention and customer service.

These strategic choices each have a name, coined by Michael Treacy and Fred Wiersema (1995). The first strategy is called *operational excellence*. With this strategy, producers pay particular attention to costs and are excellent at streamlining processes. Treacy and Wiersema have grouped the second and third strategies under the heading of *product leadership*. They call the last strategy *customer intimacy*. According to Treacy and Wiersema, a company would do well to choose a single strategy—provided do not underperform on the other dimensions (the *threshold value*)—and to implement it throughout the company. Only then can they distinguish themselves from their competitors and make a profit (Fig. 1.1).

Professor Michael Porter (1985) also names the three strategies, only in slightly different words:

1. Cost leadership (corresponding to operational excellence): The company competes with the lowest possible costs and thus the lowest possible price.
2. Differentiation (product leadership): With unique products the company tries to distinguish itself from the competition.

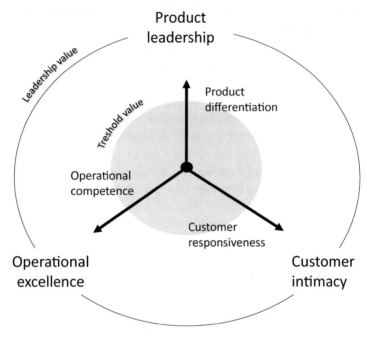

Fig. 1.1 The three strategies of Treacy and Wiersema (1995)

3. Focus (customer intimacy): The company focusses on specific consumer segments (e.g., young people or women) and builds a specific relationship with them (Table 1.1).

This trinity is also leading in European Union policy: 'competition policy encourages companies to offer consumers goods and services on the most favourable terms. It encourages *efficiency* and *innovation* and *reduces prices*'.[2]

Consciousness or not, these strategies reflect consumers' key selection criteria for purchases. Consumers prefer products that are innovative and exemplify high quality (product leadership), offered at a good price (operational excellence), and providing a strong customer relationship (customer intimacy).

Table 1.1 Firms' competitive strategies

Competing on	Treacy and Wiersema	Michael Porter
Price	Operational excellence	Cost leadership
Quality	Product leadership	Differentiation
Innovation	Product leadership	Differentiation
Customer relationship	Customer intimacy	Focus

Many people no longer buy just to survive. A society in which buying (unnecessary) products plays an important role is commonly referred to as a *consumer society* (Image 1.3).

Image 1.3 Consumer society (*Source* Pixabay, StockSnap)

For some, the meaning of life lies in the constant search for material things to buy. You buy, so you exist. You are bored, so you buy. Your identity is defined by what you buy: you are cool, sporty, fast, young, hot and rich. You are unhappy, so you spoil yourself with something nice. Buying gives short gratification and can even be addictive. A student once said that one day she became aware of her own urge to buy and decided to keep her wallet closed for a month. Only then did she notice how addicted she was to buying. Her ban made her restless at first, but then she calmed down.

Therefore, companies are challenged by consumers to invest in innovation, high quality, or reducing costs. This provides the consumer with the best products at the best price. But in that challenge lies also the *compelling hand of the free market*. Goudzwaard and De Lange (1991) formulate this dilemma in *Genoeg van te veel, genoeg van te weinig* as follows: 'A company that decides to take greater care of the surrounding environment, the quality of work and the preservation of as many jobs as possible would quickly price itself out of the market because of the costs involved, and go under economically'. A little further on: 'these are indications that we are dealing here with a compelling influence of an (economic) social order' (pp. 74 and 75).

The free market ensures that companies (have to) make choices that are primarily focussed on what the customer wants. Innovation, price, quality and customer relations should lead to profit at the end of the year, only then companies keep their competitive position. The free market, with so many suppliers, is in essence *survival of the fittest*.

Example

In the early 1990s, the mobile phone appeared on the market. It was a super-innovation that made calling possible everywhere. The Finnish company Nokia played a leading role in this new market. As a front runner, they had a lead on their competitors and recorded high profits. However, when Apple introduced the mobile phone with touchscreen, consumers switched en masse, which led to a profound crisis at Nokia. Supply and demand: if the competition is better and smarter, your customers will disappear like snow in summer. This is described as 'the coldness of the market': it is only about the properties of the product, not about other aspects, such as who made it or how it was created (Image 1.4).

Image 1.4 An old model of Nokia (*Source* Kostkarubika005)

Consumers therefore play an important role in the market: their choices influence the strategies of companies. But they are not the only player. In recent decades, shareholders have also emerged as an increasingly powerful player, influencing companies' strategic choices. You will learn about their role in the next section.

Shareholders strive for high returns on their investments

Individuals and investors, such as insurers and pension funds, are looking for ways to grow their assets, especially when a savings account at a bank hardly yields any interest. Another way is to invest the money in shares of companies. With a share, someone becomes co-owner of the company. The market where these products are sold is called the *financial market* or the *stock exchange*. Most investors look for stocks that they expect to pay the most dividends. Dividends are a portion of a company's profits paid out to shareholders. *The more profits a company makes, the more dividends* can be paid out to shareholders (companies may also reserve a portion of profits for investment in the company). Shareholders therefore are incentivized to encourage companies to make as much profit as possible.

A second benefit for shareholders is the potential for price gains on their shares. Price gain refers to the difference between the price at which the shareholder initially purchased the shares and the price at which they sell them. When a company demonstrates success and profitability, the demand for its shares tends to increase. If the company does not issue new shares, this heightened demand can drive up the share price. As a result, shareholders can realize a profit by selling their shares at a higher price than they initially paid. Shareholders keep a close eye on these prices to decide when to sell their shares. Their share is therefore also a means of speculation. From this perspective too, shareholders find more profit attractive: good company results lead to increased demand for shares and higher share prices. Falling profits are often immediately punished by a quick sale of shares, before the price takes a dramatic fall. Shareholders and top management therefore focus on increasing profits.

Third, high profits help the company to gain *access to more capital*. Since increasing profits means more people want to buy its shares, a company can issue extra shares. Additionally, banks are willing to provide loans to companies with good profit expectations.

Lastly, high profits help in the *overall sales of the business*. Your business has then become your product.

So there are four reasons why maximum profit is attractive both for the company and for shareholders and investors:

1. The more profits, the more dividends for shareholders.
2. The more profits, the more the share price will rise.
3. The more profits, the more access the company has to capital.
4. The more profits, the easier it is to sell the company at a high price.

PE firms

Some companies' business model is acquiring companies with high-profit expectations; these are so-called private equity firms, or PE firms for short. PE firms buy shares in a company to make a financial gain. Eileen Appelbaum and Rosemary Batt, two scientists who conducted in-depth research into the practice of private equity, distinguish two categories of PE firms (or rather, they speak of 'two faces' of PE).[3]

The first category acquires financially distressed companies that are often overlooked by other investors. The PE firm invests in it, reorganizes it and—if successful—turns it into a thriving business. They take losses if they don't succeed in fixing up the company, but make a profit if they do. Such PE firms often have a portfolio of companies in which they invest. For the company being bought out, it can also be a boon because PE firms often have a lot of knowledge on how to make a company profitable.

The second category of PE firms focus on profit maximization for their directors, and are less interested in the acquired company. One strategy employed by these firms is to acquire struggling companies that possess significant real estate assets. They secure funds from institutional investors who aim to expand their asset base through the private equity firm's fund. The private equity firm itself typically contributes only around 2% of the investment. Once in control of the company, they separate the real estate portion and often sell it for a substantial profit. In this way the PE firm can quickly convert its invested money back into liquid cash. The bought-out company then has to pay rent for the property it used to own. So the PE firm does not use the money earned from the sale of the real estate to make improvements in the purchased company. Because the problem is not solved by selling the real estate, the PE firm will also enact cost savings in the company, will charge consulting fees for additional advisory work and receives a commission for its work for the institutional investor.

The concept of shareholder value has long been central to economics. Nobel Prize winner for Economics Milton Friedman said in 1962 that companies should exist primarily to serve their shareholders.[4] The concepts of profit and profit maximization are still central to companies today. However, a narrow focus on serving the interests of one group often results in the neglect of other stakeholders. For instance, when making investment decisions, if the primary consideration is maximizing profit without due regard for the increased workload it may impose on employees, the company may proceed with the investment. Similarly, in pursuit of cost reduction, companies may opt for manufacturing in regions with lower labor costs, even if it involves the exploitation of child labor. In such cases, profit takes precedence over other ethical and social considerations (Image 1.5).

Image 1.5 The stock market is an attractive but risky way to increase your capital (*Source* Pix1861)

Some change is occurring. The Dutch columnist Stevo Akkermans writes:

In mid-August 2019, the top leaders of the world's 181 largest corporations, united in the US Business Roundtable, came out with a statement that was downright revolutionary. They said goodbye to the longstanding principle that 'companies exist primarily to serve their shareholders'. Instead, they now recognized that companies have 'fundamental obligations to all their stakeholders'.[5]

Not all companies are listed on the stock exchange. Worldwide there are many small and medium enterprises (SMEs), which are often family businesses. These companies do not always focus on profit maximization, but they too are subject to the coercive hand of the free market and have to maintain their competitive position.

The Dutch East India Company (VOC) was the first organization with shares[6] (Image 1.6).

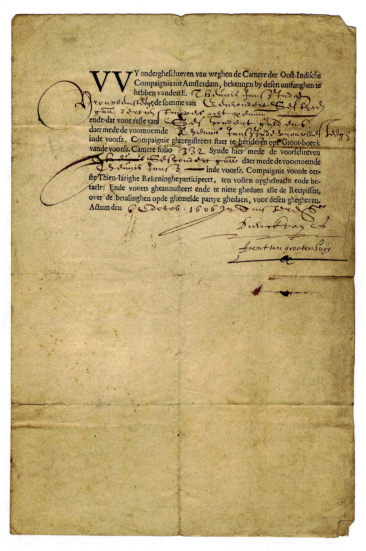

Image 1.6 One of the oldest shares of the VOC, dated 1606, in the name of the grocer Theunis Jansz for six hundred guilders (*Source* Pixabay Amsterdam City Archives)

Customers and shareholders therefore play an important role in the choices companies make to survive in the free market. In the next section, you will learn what those choices are, and then in Chapter 2 we will show that these choices hinder the transition towards a sustainable world (Fig. 1.2).

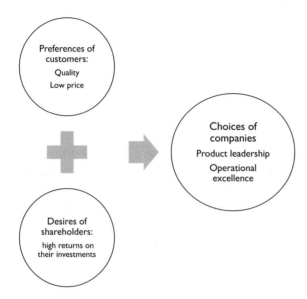

Fig. 1.2 Customer and shareholder influence corporate choices

1.2 Effects of the Compelling Hand on Business Choices

Consumers and shareholders thus force companies to deliver the most competitive product with the highest profit. As discussed in the previous section, companies have the choice to respond in different ways to this compelling hand. In this section, we will explore the options available to companies, which can be categorized into two main directions:
1. Product leadership through product improvement and innovation. This includes the strategies of innovation, quality improvement and customer relationships.
2. Excellent business processes through cost savings (and thus profit maximization). This is about efficiency and can be applied to both employees and production resources (such as raw materials, machines and transport).

Investing in Product Leadership

Product innovation and quality improvement are important drivers of economic growth. By coming up with innovative products, producers entice their consumers to make purchases. 'You believed that your bearskin would provide the best warmth, but let me introduce you to my bearskin with an insulating coating; buy mine!'

An in-house research and development (R&D) department is a traditional approach for companies to enhance and innovate their products. In the past, these departments were often closed-off and secretive, consisting of internal scientists. However, in today's landscape, they have transformed into more collaborative entities that actively engage with customers, universities, and other stakeholders. This collaborative approach fosters the development of marketable products. Product development is a time-consuming process that often spans several years, from the initial idea to the final market-ready product. This journey requires significant upfront investment and entails inherent risks. Companies are willing to make such investments only when there is a high probability of recovering the costs and generating substantial returns. Therefore, the *return on investment* (*ROI*) should be as high as possible. This also applies to a second way to innovate: *buying start-ups* whose innovations threaten your competitive position. Companies may acquire promising organizations to eliminate competition and capitalize on the potential of their new products. This strategic move allows them to leverage the acquired innovations for their own benefit. For the owner of the start-up, such an acquisition can be an enticing opportunity to achieve substantial financial gains in a relatively short period of time.

Innovations can be protected with a patent. If you have received a *patent right*, then you have an exclusive right to an invention, technical product or process. Others are not allowed to copy, sell or import your product for a certain period of time (maximum 20 years). Patent law states three requirements for obtaining a patent: novelty, inventiveness and industrial applicability. Patent protection can be arranged worldwide via the Patent Cooperation Treaty[7] (Image 1.7).

1 The Functioning of the Free Market

Image 1.7 Rusk with a notch. The notch on the side of the rusk, where exactly your finger fits in, was also granted patent protection (*Source* Own image)

With improved *quality* at a competitive price, companies can also attract new customers. Quality is a broad concept and can encompass many dimensions. In general, a product has quality if it meets the general standards that apply to that category of products. For example, headphones have quality if they produce good sound and is comfortable to wear. A quality winter coat keeps you warm when it freezes. Quality is often mentioned in the same breath as price: value for money. Consumers sometimes settle for lesser quality if the price is lower. Conversely, consumers sometimes pay a high price for products that are known for their high quality.

To deliver quality, the company's processes must be effective. Various tools and guidelines have been developed to support companies in this process. A well-known quality standard is ISO 9001 of the International Standardization Organization.

Reduced quality can however also be an innovation and therefore the engine for new sales. For example, making fabric that deteriorates faster means customers will have to replace it more often. Both improved and reduced quality leads to continuous replacement of old products.

Marketing plays a vital role in product innovation and improvement as it encompasses the actions taken by a company to raise awareness and promote its products to customers, including advertising and other promotional activities. While individuals may generate new product ideas and oversee their production, the success of a product hinges on its ability to reach the target market and appeal to consumers. Effective marketing ensures that customers are aware of the product's existence and perceive it as appealing enough to make a purchase. This is why large companies in particular invest a lot of money in marketing. For example, Unilever's annual report shows that they invested around 6 billion euros in branding and marketing in 2021. Imagine what more could have been achieved with that money.

> Companies are increasingly advertising their products through influencers on social media. For example, if you have a lot of followers on Instagram with an associated YouTube channel, you may be asked by companies to unbox, wear or use their products. One of those influencers is Nikkie. With her beauty account *Nikkietutorials*, 16.4 million followers in 2022, she advertises well-known make-up brands. The downside of this method was highlighted in the documentary Follow Me,[8] which shows how Instagrammers do everything they can to gain more followers. For some it has become a day job; others buy fake followers that can be generated with software. It is also possible to automatically leave comments on posts with certain hashtags. The documentary Follow Me shows that in Russia women pay to post comments.

Marketing enables companies to familiarize customers with their strategic choices. One well-known framework in the marketing field is the four Ps: price, promotion, place, and product. Over time, additional elements such as presentation, partners, and personnel have been included, collectively forming the *marketing mix*. Through the manipulation of these elements, companies translate their strategic decisions into actionable marketing strategies. For example, marketing through social media falls under the promotion element, involving targeted advertising on specific channels to reach a particular audience. Another example is the utilization of personnel to cultivate personalized customer relationships.

Excellent Business Processes Through Cost Savings

Consumers are constantly weighing up quality against price. Competition forces companies to produce efficiently, that is, to keep costs as low as possible in order to make the highest profit at the end of the year. Without profit, there is no money for innovation and investment in improvements. Therefore, efficiency is an important guiding principle in many companies.

Companies can achieve cost reduction and efficiency in three ways:

- Companies deploy the means of production (raw materials and (raw materials and machinery) as efficiently as possible.
- Companies cut back on staff costs by having fewer people do more work (productivity).
- Companies cut back on staff costs by offering the lowest possible salary.

Cost Reduction on Resources

Take a look at your phone. Do you know what raw materials are used in it? You might be able to think of the most common ones. But did you know that your mobile phone is made up of up to forty different raw materials?[9] Important raw materials include minerals such as iron, aluminium, gold, salt and chalk. Natural gas, oil, sun, wind and coal are important sources of energy, needed to produce the phone. Raw materials also include the basis of our food supply: grain, cocoa, coffee beans, milk and the like. Clothing contains cotton, flax (for linen) and paint. And of course, water is an important raw material.

The costs of raw materials can be kept low by using less material. For many raw materials, the less companies use, the better it is for our planet. Chapter 5, therefore, examines ways of reducing the consumption of raw materials.

> The OECD[10] shows that 'the general trend in OECD countries is towards higher material productivity and lower per capita consumption of materials. But levels of material consumption per capita remain higher compared to other world regions. The material footprint, including materials extracted abroad and embodied in international trade, has increased in many OECD countries' (Image 1.8).

Image 1.8 Mining for raw materials such as copper and lythium leaves deep scars in nature (*Source* Pixabay, Shibang)

However, cost savings can also be achieved by optimizing the procurement of resources. This involves minimizing the purchase price or finding more cost-effective alternatives. Additionally, companies can focus on reducing production costs. There are several strategies that can be employed to achieve these goals.

To minimize costs for certain raw materials, companies may seek to obtain them at *the lowest possible price*. In some cases, this can lead to sourcing from countries where child labor is still prevalent, such as in mines. A notable example is the Democratic Republic of Congo, which is a significant source of cobalt, a crucial raw material used in various technological devices like mobile phones and batteries.

Technological innovations have significantly improved the efficiency of resource extraction and utilization over the years. The introduction of tractors, for example, has allowed farmers to cultivate and harvest larger areas of land more efficiently compared to manual or horse-drawn methods. This *scalability* has resulted in increased food production, which is crucial to meet the demands of a growing global population. Moreover, it has contributed to reduced production costs per hectare, as fewer farmers are now required to operate large machinery. However, with the expansion of agricultural activities, the maintenance of larger land areas has become a challenge. Manual weed and pest control on such vast scales are impractical, prompting many farmers to rely on pesticides. These chemical substances help protect crops from pests, insects, and fungal infections, as well as inhibit the growth of weeds that compete with cultivated plants. The use of pesticides can be seen as a response to the need for efficient crop protection and maintenance.

Meat production also shows cost saving by producing on a larger scale. This has led to the development of factories in which animals are caged close together. Around two-thirds of farm animals worldwide are factory farmed every year—that's nearly 50 billion animals.[11] Meat production has become more cost-effective due to economies of scale, particularly in the context of factory farming. By confining animals to smaller spaces, less land and energy are required to maintain them. Moreover, animals in factory farms are fed concentrated diets, which promote rapid growth and reduce processing times. For instance, a broiler chicken bred for meat consumption can reach a weight of 2 kg in just six weeks, making it ready for slaughter. In contrast, chickens raised under natural conditions take at least twice as long to reach the same weight before they are processed. The website of Compassion in World Farming provides more information on this topic.

In recent decades, significant advancements in fishing methods have been made, particularly in the realm of trapping fish. New technologies such as echolocation and the pulse method have revolutionized the fishing industry.Echolocation involves the use of specialized equipment to detect and locate schools of fish underwater through the transmission and reception of sound waves. The pulse method, on the other hand, utilizes electric shocks to startle fish, causing them to swim into the waiting nets.

Economies of scale also play a significant role in cost reduction in the aircraft industry. An example of this is the transition from smaller aircraft to larger ones, such as the introduction of the Boeing 747 in the 1970s. With the capacity to carry significantly more passengers compared to its predecessors, the Boeing 747 achieved a tenfold increase in passenger capacity per flight. This increase in scale resulted in substantial cost savings per passenger, making air travel more affordable and accessible to a larger number of people.Similarly, sea transport has also witnessed improvements in efficiency through the utilization of economies of scale. By maximizing the number of containers loaded onto cargo ships, sea transport became much more efficient. The ability to transport larger volumes of goods per ship significantly reduced transportation costs (Image 1.9).

In addition to scaling through technology, you can also *cheat consumers*, allowing you to keep costs down. For example, in 2013 the Dutch Food and Consumer Product Safety Authority discovered that horse meat was being mixed with beef, even though it was sold as pure beef. Customers can also be fooled by sustainability labels. Anyone can design their label and use it to promote their products as sustainable. Chapter 2 discusses the consequences of this in more detail.

Finally, you can keep the costs down by using *clever tax structures*. The profits that companies make are taxed in the country where the company (or subsidiary) is managed. Each country determines its corporate tax rates.

Image 1.9 A Boeing 747-400 holds 408 passengers (*Source* Pixabay, PatrickE)

Companies, therefore, work with advisers to devise business structures that allow them to continue to work and live in a country with a high tax rate but pay tax in a country with a low rate. The countries with low tax rates are known as tax havens. Well-known countries where many companies are supposedly based are Bermuda, the Cayman Islands and The Netherlands. For example, the pop group The Rolling Stones kept its headquarters (no more than an address) in The Netherlands for a long time.

Managing Productivity of Employees

In addition to saving on the cost of resources, companies can also save on the cost of employing people. First, more production per person (and therefore lower costs per product) is achieved by focussing on productivity. By making employees work harder, companies need fewer of them. Within our society, the prevailing emphasis on performance and productivity has given rise to the concept of a "performance society," wherein individuals are viewed as valuable resources required to achieve desired outcomes and generate financial gains.

A common way to manage this productivity is by making agreements or by *setting goals* that an employee has to achieve in a quarter or a year. This can be anything: delivery times, turnover for clothing sellers, number of consultations for psychiatrists or number of publications for scientists. Sometimes financial rewards are linked to this. For example, a salesperson can earn more money the more they sell, and a senior lecturer can advance to the position of professor through promotion, primarily based on their exceptional track record of producing substantial and high-quality publications. But it also works the other way around: if an employee doesn't meet their target, demotion or even dismissal can follow. These *incentives* motivate people to do their best, because a reward awaits them or because they fear dismissal. This way of managing fits well in the market economy, which is based on self-interest. Setting individual goals focusses the person on themselves and not on what is needed collectively, according to Vosselman and De Loo (2018). Less or no compensation for overtime or giving someone more tasks for the same number of hours also falls under the heading of productivity.

For people working on an automated production line, *standardization or specialization of actions* leads to more production. Adam Smith (1723–1790) saw how specialization could lead to increased productivity. Smith lived in the time of industrialization, in which machines could facilitate the work of humans, but also take over. By requiring people to perform only one action, a routine is created, enabling people to perform that action faster. Factories can then even measure the optimal speed and adjust the conveyor belt so that employees must achieve that speed. Frederick Taylor (1856–1915) extensively researched the concept of specialization, which subsequently became known as *scientific management*. Nowadays there is a button at the conveyor belt so that people can stop the line.

Automation and *robotization* are increasingly leading to replace human employees. Whereas in the past manned pallet trucks transported goods through the warehouse, now fully automated computer-controlled conveyor belts perform this function at the larger online stores. Because consumers now do a lot of banking online, far fewer bank employees are needed. A study by the Central Planning Bureau and others shows that about 8% of employees lost their jobs due to automation.[12] The Organisation for Economic Cooperation and Development (OECD)[13] found that 'on average across the 21 OECD countries, 9% of jobs are automatable'.

The *merger* of companies can also increase productivity, for example, by reducing purchasing and sales costs. Now a single buyer can buy supplies for multiple companies. This increases the productivity of this employee and provides a lower cost for buying larger quantities. In addition, companies merge because one company gains access to the unique characteristics of the other company. For example, a company may gain access to markets in which the merger partner operates, to new technology or to better management. This saves both company's development costs (partly consisting of personnel costs).

Cuts in Salary Costs

Cost reduction through keeping salaries low is another approach that companies may adopt. There are several strategies to achieve this. First, many companies work with *job profiles with salary scales*, where people with little experience start at the low end of the scale. In our society, many people think people with the least education should receive the lowest salary. But is this true? Cleaners and garbage collectors, for example, perform a very important task in our society. This becomes clear when they go on strike. They work hard: starting work at 5 a.m. is normal. And yet they earn just over the minimum wage. The main reason for a difference in wages is that also in this (labour) market, the law of supply and demand determines the price. Many people can clean, and few understand the software of the TV boxes. So the technician can negotiate a higher wage than the cleaner. On the other hand, you often see that the top of the company gets richer and richer.

> **The richest 1% bagged 82% of wealth created last year—the poorest half of humanity got nothing** For years, Oxfam Novib has been working on inequality between people. In their press release they state the following:
>
> > We are living through an unprecedented moment of multiple crises. Tens of millions more people are facing hunger. Hundreds of millions more face impossible rises in the cost of basic goods or heating their homes. Poverty has increased for the first time in 25 years. At the same time, these multiple crises all have winners. The very richest have become dramatically richer and corporate profits have hit record highs, driving an explosion of inequality.
> >
> > - Since 2020, the richest 1% have captured almost two-thirds of all new wealth—nearly twice as much money as the bottom 99% of the world's population.
> > - Billionaire fortunes are increasing by $2.7bn a day, even as inflation outpaces the wages of at least 1.7 billion workers, more than the population of India.
>
> *Source* Oxfam International, 16th of January 2023[14]

Companies must be able to *respond flexibly* to changing customer demand. If demand increases, companies need more employees. If the economy goes into recession and demand drops, a company has to shrink its workforce. However, an employer cannot simply dismiss their employees. They must comply with the rules of dismissal law, which can be expensive. To avoid these costs, employers often make use of flexible labour. The OECD distinguishes two types of flexible employment relationships: temporary employment and self-employment. In European countries, about 14% of workers are temporary and 15% are self-employed.[15] If the group for a particular service is large, companies can choose people who agree to the lowest hourly rate (after all, in this market too, supply and demand determine the price). This hourly rate is often much lower than the hourly wage that companies would have to pay their employees. Costs such as sick leave, holiday allowance and pensions are included in the employee hourly rate, but often not in the external hourly rate. Independent contractors who want to include these costs in their rates lose work to people who offer their services at a lower hourly rate. In an industry where there is little supply of freelancers or a profession where rates are historically high, this does not apply. A good example is lawyers, who are difficult or impossible to find for rates lower than 250 euros per hour.

Wage restraint, demotion, lower pay for the same job and *dismissal* are methods employed by companies to minimize wage costs. Wage restraint involves limiting salary increases, often below the inflation rate, to control expenses. Demotion refers to offering employees lower positions within the company, often accompanied by a reduction in salary. During times of reorganization, companies may negotiate with works councils to secure lower wages to ensure the survival of the organization. Dismissing employees is a last resort to reduce labor costs, typically authorized by the courts if it is demonstrated that the company would otherwise face bankruptcy.In more extreme cases, companies may adopt radical measures to save on labor costs. For example, they may declare bankruptcy and subsequently rehire the same employees at lower wages or under less favorable employment conditions. It is important to note that these approaches can have significant implications for employees and raise ethical considerations.Please let me know if there is anything else I can assist you with.

Labour costs can also be kept low by *offshoring*. Offshoring means that the activity previously carried out by the company itself is moved abroad, while the head office does not move. Sometimes the activity remains part of the company; this is called *captive offshoring*. If the activity is no longer part of the home company, it is called *outsource offshoring*. A 2018 study by Statistics Netherlands shows that the activities most commonly outsourced are the production of goods and administrative and management functions.[16] In all the periods studied from 2001 to 2016, saving on labour costs is the main motive for choosing offshoring.

Lastly, it is important to address the issue of *child labour*. While it is prohibited in many countries, unfortunately, it still persists in some regions. According to UNICEF, approximately 60 million children were engaged in labour worldwide at the beginning of 2020.[17] There is a good chance that you have products that were partly made by children.

1.3 Summary

Market and market forces

In the market, supply and demand balance themselves with the help of price. If the demand for a product is high, more suppliers market this product. If the demand decreases, the production decreases and companies disappear through bankruptcy.

The main players in the free market

This book distinguishes three main players: customers, who make choices about what they buy; shareholders, who decide where to invest their money; and the companies they wield influence upon.

The effect of the free market

Through market forces, companies are challenged by customers and shareholders to meet their wishes. This challenge is the compelling hand of the free market. If companies do not meet the wishes of customers and shareholders, they lose their competitive position and eventually go bankrupt.

Company strategies to meet customer and shareholder requirements

Companies apply three strategies to retain customers and shareholders. They use these strategies to maximize profits, satisfying the shareholders.

1. They focus on improving and innovating their product (product leadership/differentiation).
2. They build an strong relationship with customers by investing in customer attention and customer service (customer intimacy/focus).

3. They lower their prices by optimizing their business processes (operational excellence/cost leadership). This often means cutting costs to remain profitable.

Product improvement, innovation and customer relationship in practice

Companies work on innovation, quality improvement and customer relationships in the following ways:

- investing in research and development
- acquisition of innovative start-ups
- patenting applications
- increase quality (or deliver less quality)
- creating a good customer relationship by using the marketing mix.

Ways to keep costs down

Companies can keep their costs down by saving on the consumption and price of resources and employees.

The cost of consuming resources can be kept low by:

- putting pressure on the purchase price
- more efficient use through, among other things, economies of scale
- improved and smart use of technology
- fraud
- tax structures.

Labour costs can be kept low by:

1. managing employee productivity:
    - setting targets
    - standardizing work
    - automation
    - merging activities
2. keeping salary costs down:
    - using job profiles
    - wage moderation, demotion, lower pay for the same job and dismissal
    - working with flexible labour
    - offshoring
    - child labour

Notes

1. 13 biggest environmental problems of 2022. Accessed September 30, 2022, from https://earth.org/the-biggest-environmental-problems-of-our-lifetime/.
2. Competition policy. Accessed September 30, 2022, from https://competition-policy.ec.europa.eu/index_en.
3. Book talk: Private equity at work: When Wall Street manages main street. Accessed September 30, 2022, from https://youtu.be/WMty16bIYhw.
4. The social responsibility of business is to increase ... what exactly? Accessed September 30, 2022, from https://hbr.org/2012/04/you-might-disagree-with-milton.
5. Will capitalism ever get a human face? Accessed September 30, 2022, from www.trouw.nl/opinie/krijgt-het-kapitalisme-ooit-een-menselijk-gezicht~bdd25623/.
6. Special VOC share from 1606 discovered. Consulted on September 30, 2022, from https://nos.nl/artikel/2060982-bijzonder-voc-aandeel-uit-1606-ontdekt.html.
7. Worldwide patent applications can be made via WIPO, the global forum for intellectual property (IP) services, policy, information and cooperation. Accessed August 22, 2022, from https://www.wipo.int/portal/en/.
8. The first documentary about Instagram on Instagram. Accessed August 23, 2022, from https://www.instagram.com/followme.doc/.
9. European Commission proposes new strategy to address EU critical needs for raw materials. Accessed September 30, 2022, from http://europa.eu/rapid/press-release_IP-08-1628_en.htm.
10. OECD material consumption. Accessed August 23, 2022, from https://www.oecd-ilibrary.org/sites/f5670a8d-en/index.html?itemId=/content/component/f5670a8d-en#section-d1e5457.
11. Factory farmed animals. Accessed August 25, 2022, from https://www.ciwf.org.uk/factory-farming/animal-cruelty/.
12. Automatic reaction. Accessed August 25, 2022, from https://www.cpb.nl/sites/default/files/omnidownload/CPB-Discussion-Paper-390-Automatic-Reaction-What-Happens-to-Workers-at-Firms-that-Automate.pdf.
13. The risk of automation for jobs in OECD countries. Accessed August 25, 2022, from www.oecd-ilibrary.org/social-issues-migration-health/the-risk-of-automation-for-jobs-in-oecd-countries_5jlz9h56dvq7-en.
14. How we must tax the super-rich now to fight inequality. Accessed January 20, 2023, from https://www.oxfam.org/en/research/survival-richest.
15. Self-employment rate. Accessed September 1, 2022, from https://data.oecd.org/emp/self-employment-rate.htm#indicator-chart.
16. Offshoring. Accessed September 1, 2022, from https://www.cbs.nl/en-gb/news/2018/26/over-30-thousand-jobs-offshored.
17. Child labor in the world. https://www.unicef.org/protection/child-labour.

Literature

Goudzwaard, B., & de Lange, H. M. (1991). *Enough of too much, enough of too little*. Ten Have.

Porter, M. (1985). *Competitive advantage: Creating and sustaining superior performance*. Free Press.

Treacy, M., & Wiersema, F. (1995). *The discipline of market leaders*. Addison-Wesley.

Vosselman, E., & de Loo, I. (2018). Performativity in networks: The Janus head of accounting. *Monthly Journal of Accountancy and Business Economics, 92*(1/2), 21–25.

2

The Tragedy of the Compelling Hand

In the previous chapter, you learned about market forces and how companies develop, maintain and strengthen their competitive position in the market. You also know that the compelling hand of the market forces companies to make these choices. When consumers want innovative products with high quality for a low price, and shareholders strive to maximize profits, companies will offer their products and services to meet these demands.

However, the essence of this narrative lies in the impact of market forces: the decisions made by companies often result in significant negative effects, commonly referred to as negative externalities. Well-known examples are climate change, child labour and burnout. It is also the conclusion of Bas van Bavel (2018) in his book '*The Invisible Hand?*': market economies erode prosperity, equality and broad decision-making within a society. This chapter shows the harmful effects each choice produces for both people and the environment.

2.1 Damage Caused by Product Leadership

First, you will learn about the damage caused by the production of goods in general. How our increased consumption and production, fuelled by product innovation and quality improvement and customer relationships with associated marketing can lead to irreparable devastation is well illustrated in the famous *tragedy of the commons*, first described in 1833 and still relevant today. The story goes like this:

> Once upon a time there was a pond with ten fish. Around the pond lived five families, each of which needed one fish a day to survive. The ten fish together produced five fish a night. So if the community of five families wanted to survive, they had to eat no more than one fish a day. That way there would be enough fish in the pond to eat every morning. The families could not see each other while fishing. So they could not see how many fish the others took out of the pond in the morning. If someone would only take two fish from the pond once, there would only be four new fish in the pond the next day, because one fish would not have had a partner to produce young fish. So one family would then catch one old fish, leaving only eight. This pattern would repeat itself until all the fish were gone for everyone.

Let this story sink in. It is in everyone's *long-term* interest to get just one fish out of the pond each day. However, *in the short term* it is very attractive to catch more fish to have more for your own family at that moment. But this is also bad for yourself in the long run. That is the *tragedy of the commons*: it is a story about the overexploitation of a common asset (the *commons*), where everyone knows that if everyone does it, it will eventually run out. But since everyone else is probably doing it, and it's hard to trust that the other person will keep their end of the bargain, you do it too. That way, you benefit yourself in the short term.

> **Tip**
> The tragedy of the commons is a now familiar story, and it sounds almost like a play. The idea was developed in an article by ecologist Garrett Hardin (1915–2003), who was concerned about the growing world population and production of food (Hardin, 1968). He based his reasoning on the story of the commons, written down in 1833 by William Forrester Lloyd (1794–1852). Hardin's entire story is available in pdf on the internet.

Can you already translate this story to our current society?

In Chapter 1 you learned about the characteristics of the market. This market works because everyone acts out of self-interest and aims at a profit for themselves. Consumers buy products to benefit themselves: the products help to survive and make life more pleasant. Consumers are tempted to buy products over and over again. Companies also offer products to earn money and—encouraged by shareholders—make a profit. The principle of profit maximization is also the principle of self-interest maximization. The functioning of the market can be seen as a real-life example of the Tragedy of the Commons.

One can apply the concept of the tragedy of the commons to air travel. Air travel has become increasingly accessible and affordable, allowing people to explore new destinations and experience different cultures. However, this progress comes at a cost. Airlines strive for maximum profitability by operating more flights and increasing occupancy rates, often relying on tax-free kerosene and efficient aircraft. In the short term, this benefits both travelers and the airline industry, creating jobs and providing affordable travel options. However, the tragedy of the commons reminds us that actions driven by individual interests can have long-term consequences for society. Air travel contributes significantly to CO_2 emissions, which in turn contribute to climate change, leading to severe consequences such as natural disasters and rising sea levels. The dilemma arises when individuals recognize the environmental impact of flying but feel that their personal choices won't make a difference if others continue to fly. This collective action problem leads to a cycle where individuals continue to fly because everyone else does, perpetuating the harmful effects on the environment.

Flying is just one of the many products we purchase. Have you ever thought about the path our products take? The raw materials are mined, water is consumed, groups of people convert the raw materials into usable material, other people assemble all that material into our products, and they are traded, packed and transported until you exchange your money for the purchase of the product, and carelessly throw it away after using it or donate it to a thrift store. This is how we live. The described process of extraction, use and disposal of raw materials is called *linear use*. Figure 2.1 shows how this is often represented.

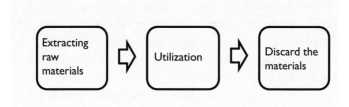

Fig. 2.1 Linear use of resources

The earth is our common. In 1987, the now famous Brundtland report was published, named after the Norwegian prime minister of that time. The report, which is officially called *Our common future*, describes the relationship between economic growth, the ecological damage this growth causes and the relationship with poverty.

Damage arises from the linear use of raw materials and has three categories:

- Overexploitation: we consume more than the earth can give us.
- Damage to our environment (humans, animals and the earth).
- Waste: the products are disposed of as waste after use.

Additionally, innovations are not available to everyone. This is due to a focus on high return on investment and patent rights.

Overexploitation

Human consumption is exceeding the Earth's regenerative capacity, resulting in an unsustainable depletion of resources. For example, a forest needs recovery time after it has been cut down, fish must first have young before they can be fished again, and coal simply runs out at some point (it is not clear whether all raw materials will run out, opinions differ on this). In order to raise awareness about this ecological debt, Wackernagel and Rees, through their organization the Global Footprint Network, introduced the concept of "overshoot day." This is the specific day when the Earth's available resources are exhausted and the planet is subjected to overexploitation. This day changes every year. In 2022, that day was on the 28th of July, a deterioration compared to the previous year, when the day fell on 30 July (after all, the longer people can live with the available resources, the better). Only in 2020 was there an improvement, which was due to the corona pandemic. Figure 2.2 shows how this day falls earlier and earlier in the year.

The ecological footprint varies from country to country. For example, some countries in Africa consume less than one globe. This may be due to an abundance of resources or low consumption of resources. Hence, the market, where people engage in the exchange of products, contributes to an excessive strain on the ecosystem as it consumes and depletes valuable resources.

Overshoot day is one way of making visible where the boundaries of our earth are. Another model is the Planetary Boundaries overview of the Stockholm Resilience Centre. This overview contains nine planetary boundaries, of which, according to them, four have already been crossed as a result of our actions (Fig. 2.3).

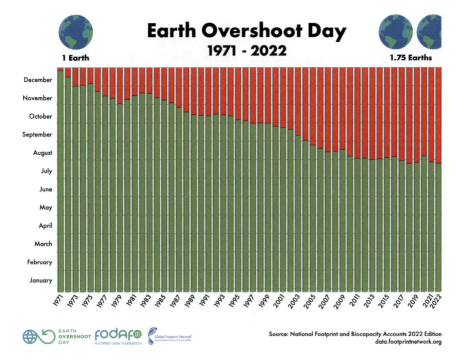

Fig. 2.2 Overview of the date when as many of the Earth's resources were used as the Earth can give (*Source* Global Footprint Network[1])

A major cause of this overexploitation is that our consumption has increased. Needs are finite, but desires are not. If your friend has the latest mobile phone, it affects your appreciation of your old one. The fact that human beings desire the same things as others is called *mimetic desire*. Marketing exploits this by causing people to develop feelings of inadequacy that can be filled by buying the product (or so advertising tells us). For example, people with mobile phones are seduced into buying a new one by the better quality of the screen or an upgrade of the camera. Software updates no longer work on the old models. You buy wireless earphones, but then you are still tempted to buy an extra wire because otherwise the earphones get lost. Fashions change, and so do the contents of your wardrobe. After a few years, the furniture isn't hip anymore, the TV can be even bigger and the bed you bought three years ago squeaks a bit too much and needs to be replaced by a better version. Increasing purchasing power has supported this rise in consumption.

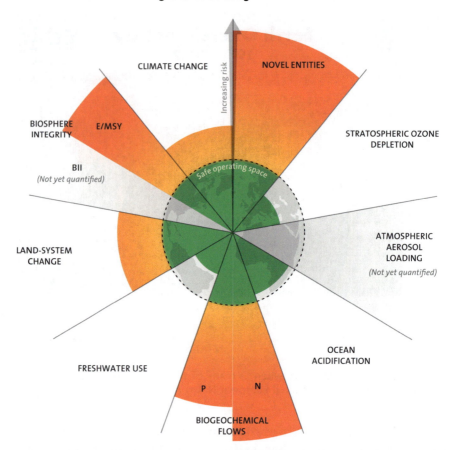

Fig. 2.3 The Planetary Boundaries overview, designed by Azote for Stockholm Resilience Centre (*Source* Stockholm Resilience Centre, CC BY 4.0)

Damage to Humans, Animals and the Earth

In addition to overexploitation, the extraction and consumption of raw materials inflict various other forms of damage. For example, the extraction and use of our main fossil energy sources—oil, gas and coal—has led to one of the biggest environmental problems of this century: *climate change*. Burning these resources releases CO_2, which most scientists now say is causing global warming (other greenhouse gases, such as methane and nitrous oxide, also contribute. These gases are converted into CO_2). This warming has been going on for some time. NASA shows that the average temperature worldwide has increased by 1.01 degrees since 1880.[2]

Rising CO_2 emissions are partly related to the growth of industry, which uses much more polluting energy sources than previous production methods.

The steam engine required coal, later replaced by the use of cleaner gas. But that too leads to emissions of CO_2. In addition, the industrial era has led to an increase in prosperity, with corresponding consumption patterns in many parts of the world. In combination with a growing world population, this increased spending leads to continuous consumption of raw materials and a continuous flow of transport. This means more ships overseas, more trucks on their way to warehouses, and more delivery vans with internet orders on their way to your home.

In 2019 total goods transport activities in the EU-27 are estimated to amount to 3392 billion tkm (one tonne of goods over a distance of one kilometre; 2018: 3353). This figure includes intra-EU air and sea transport but not transport activities between the EU and the rest of the world. Road transport accounted for 52% of this total, rail for 12%, inland waterways for 4.1% and oil pipelines for 3%. Intra-EU maritime transport was the second most important mode with a share of 28.9% while intra-EU air transport only accounted for 0.1% of the total.[3]

The consequences of climate change are major. The global sea level has risen by 1 cm since 1995 and 'Arctic sea ice is now shrinking at a rate of 13% per decade, compared to its average extent during the period from 1981 to 2010'.[4] Coral reefs are disappearing because the water is getting warmer. Already islands in the Pacific Ocean are disappearing.

Tip
The former president of Kiribati, Anote Tong, fought for years to preserve his islands, but in vain. A documentary, to be found on the internet, has been made about his struggle, sorrow and surrender.[5]

The weather is also becoming more extreme. In many countries, spring starts noticeably earlier every year, and torrential rainfall and longer periods of drought are becoming more frequent. The intensity of hurricanes is also increasing. In some parts of the world, daytime temperatures can exceed 45 °C. Old records are also being broken every year. For example, in the summer of 2022 the temperature in Coningsby, UK reached a record high of 40.3 °C. Bloomberg reports that 'new data from the U.S. government's temperature takers just ranked 2021 as the sixth-hottest year, near the very top of a list that stretches back into the 19th century. The hottest eight entries in the federal heat records have all occurred in the last eight years'.[6]

This global problem is now high on many political agendas. Fortunately, alternative energy sources exist, such as nuclear, solar and wind (see Chapter 5). However, the transition to these sources is hindered by economic, social and technical issues:

- Economic: stopping polluting energy sources means a reduction in income for the government (it receives income for the gas) and the disappearance of jobs in, for example, coal-fired power stations.
- Socially, nuclear energy is a problem because of the major consequences of accidents, as in Chernobyl in 1986, and the storage of waste (see next paragraph).

Example

National Geographic reports the following about the Chernobyl disaster: 'On 25 April 1986 routine maintenance was scheduled at V.I. Lenin Nuclear Power Station's fourth reactor, and workers planned to use the downtime to test whether the reactor could still be cooled if the plant lost power. During the test, however, workers violated safety protocols and power surged inside the plant. Despite attempts to shut down the reactor entirely, another power surge caused a chain reaction of explosions inside. Finally, the nuclear core itself was exposed, spewing radioactive material into the atmosphere'[7] (Image 2.1).

Image 2.1 This is what Pripjat (a place near Chernobyl) looks like: abandoned and overgrown (*Source* Pixabay, Wendelin_Jacober)

- Furthermore, it is currently not technically feasible to power aircraft solely with clean energy sources, which means this mode of transportation continues to rely on oil production. While electric cars have been introduced into the market, their high cost remains a barrier for widespread adoption. Additionally, the availability of clean energy sources is still insufficient to meet the global energy demands adequately. There is an ongoing debate whether solar and wind energy will provide sufficient supply or whether nuclear energy will also be needed.

The consumption of raw materials also leads to *pollution and degradation of human habitat*. Some examples are:

- Oil extraction has caused significant damage in the areas where extraction took place. For example, in 2017 a study revealed that babies in Nigeria are twice as likely to die in the first month of their lives if their mothers lived near an oil leak before they became pregnant.[8] How those oil leaks occurred has been the subject of a legal battle for years: is it because locals are illegally tapping oil, or is it also due to poor maintenance of the pipelines? In addition, accidents have occurred with drilling rigs and oil tankers, resulting in oil spilling into the sea. In 2010, for example, there was an explosion on a drilling platform in the Gulf of Mexico that killed eleven people. It took almost three months before the pipeline could be properly shut down, and all that time millions of gallons of oil per day flowed into the sea! Additionally, in 2018, an independent study by the U.S. Department of Justice revealed that another pipeline had been damaged since 2004, leaking some 38,000–113,000 gallons of crude oil into the Gulf of Mexico every day.[9] Oil spills have an enormous impact on the environment: oil slicks on beaches, a sea that becomes heavily polluted, and birds that die or are covered in oil.
- In the Netherlands, inhabitants of the province of Groningen have directly witnessed the detrimental consequences of gas extraction. In 1959, one of the largest gas fields in the world was discovered in the Groningen fields near Slochteren. A time of national wealth arrived: with the gas revenues, The Netherlands financed many government expenditures each year. Unfortunately, the euphoria came to an end when the first earthquakes occurred in the 1990s. Despite increasing damage, it took until 2018 for the government to reduce gas production and it wasn't stopped completely until 2022. Such quakes may persist for years after the gas tap is turned off.

- Trees are also an important raw material for many products. They have been used for centuries to build houses and make furniture, and we burn wood to warm ourselves. Logging is currently performed on a massive scale. Increasing agricultural activities are also a major cause of deforestation. Deforestation both affects the landscape and increases the climate problem, as trees absorb a lot of CO_2 from the air. Forests are essential not only for humans but also for the animals that live in them.

> According to nature organizations such as the World Wildlife Fund and Greenpeace, almost half of the earth's original forests have already disappeared. A study (Popatov et al., 2017) shows that between 2000 and 2013, 919,000 km^2 of forest disappeared, particularly in North and South America and Africa. The Food and Agriculture Organization of the United Nations estimates that 'since 1990 420 million hectares of forest have been lost through conversion to other land uses, although the rate of deforestation has decreased over the past three decades'[10] (Image 2.2).

Image 2.2 Deforestation (*Source* Pixabay, Camera-man)

- The use of land to produce and live on has led to a *decline in biodiversity* worldwide (the variety of life on earth), as there is little space left for nature. Worldwide, 4.6 million km² of land has been cleared for production in the past forty years. The International Union for the Conservation of Nature (IUCN) maintains a Red List of Threatened Species with the status of endangered animal and plant species around the world.[11]
- The last form of damage is *animal testing*. Animals are often used during product development, particularly for medicines (the pharmaceutical industry), animal feed and livestock. The animals most commonly used in testing are mice and rats.

Waste

People not only consume a substantial amount of goods but also discard a significant portion of them. While some items find new purpose through thrift stores and secondhand markets, the majority ultimately end up in landfills. In 2018, households processed some 218,720,000 tons (218.7 billion kilograms) of household waste in Europe alone. You can find out what that waste consists of on Eurostat's website.[12] If you have ever been to a municipal waste station, you will be amazed. The bins are full of everything: chairs, fridges, TVs, and there are separate bins for goods containing harmful substances such as asbestos. In addition to households, businesses produced 2402 billion kilos in Europe in 2018. These are residues, such as metal, left over from production. The production process also generates waste in the form of polluted water and gas emissions, among others. For example, toxic substances are used in the production of clothing, which end up in rivers via waste water.

> **The fashion industry and sustainable development goals: What is the role of the UN?**
>
> The fashion or clothing industry has an often underestimated impact on the development of our planet. This $2.5 trillion dollar industry is the second-highest user of water worldwide and produces 20% of global water waste. The production of a cotton shirt requires 2,700 liters—the amount one person drinks in 2.5 years. Ten percent of global carbon emissions (CO_2) are emitted

> by the garment industry and cotton farming is responsible for 24% of insecticides and 11% of pesticides, despite using only 3% of the world's arable land. 85% of textiles end up as waste in landfills, or 21 billion tons per year.[13]
>
> *Source* UNECE (2018) (edited)

Sometimes we buy products that, upon closer examination, do not meet our expectations. With the rise of webshops, we have learned that we can order products and return them free of charge. Consumers often assume that returned clothes are reconditioned and resold in pristine condition, but returns sometimes turn out not to be in good condition and end up in the waste stream.

> In 2018, research by German broadcaster ZDF's television program Frontal 21 and German magazine Wirtschafts Woche revealed that the online store Amazon in Germany is destroying product returns en masse, and even new products. The products in question range from refrigerators and dishwashers to mattresses and furniture. An Amazon employee told journalists that tens of thousands of euros are thrown away every day. Many products that are thrown away still work perfectly well or are even still new.[14]

The lower quality you read about in Chapter 1 also leads to more waste. As a result, this has resulted in lower prices, enabling manufacturers of such products to gain a competitive edge over their sometimes pricier counterparts.

Waste also plays an important role in the choice of nuclear energy or other energy resources. This waste *can be very harmful and degrades very slowly.*

Oil is an important raw material for all our plastic products. It is the basis of plastic bottles and polyester in clothing. It is strong and flexible and serves many purposes. However, plastic takes over a hundred years to decay in nature. Much plastic has ended up in the ocean as litter, which has led to what has become known as *the plastic soup*. Due to currents, litter accumulates in some areas to form enormous floating plastic waste sites, particularly between Hawaii and the United States. But you also find plastic in rivers or on the beach, not always visible to the eye. These pieces of plastic pose a threat to the marine ecosystem and to our health. Microplastics have even been found in our blood! The Plastic Soup Foundation has created a site with lots of information about the problem and what can be done about it.[15]

The above list of harmful effects is not exhaustive but is intended to raise awareness of the most salient ecological problems. The Tragedy of the Commons emphasizes that humans should not surpass the Earth's capacity in their utilization, to prevent overexploitation. It underscores the need to maintain a balance between human consumption and the planet's resources. There is an enormous challenge ahead of us to manage our common, our earth so that future generations can also live well here. This has not been achieved to date, because it requires a collective, global agreement. One endeavor in this regard is the United Nations Climate Convention, aimed at addressing environmental challenges. However, as of 2023, the collective determination to reverse the course and trust that all parties will adhere to the agreements is still lacking.

Damage arising from a Focus on High ROI and Patent Rights

Many sustainability issues can be solved through innovation and product improvement. The lack of progress in this area is partly because companies and their investors only work on solutions that are expected to generate profits. Product innovation is driven by the expectation of the highest possible financial return on investment (ROI), as you learned in Chapter 1. An example of this impediment is the development of an electric motor for cars. It was relatively expensive compared to the petrol engine, so consumers continued to choose diesel and petrol en masse. With little or no sales, the developer does profit from the investment. This applies to many sustainable alternatives: as long as consumers continue to opt for a low price, and investors steer towards profit maximization, investing in a sustainable product rarely pays off.

In Chapter 1 you also read about the right to a patent. With a patent, the inventor of a product acquires a monopoly, which means that others may not offer an identical product. The monopolist can therefore also charge higher prices than if they faced competition. This makes good business sense; the inventor has often put a lot of time and money into developing the product and can recover the investment with this protection. But in some cases, patents also harm society, for example, in the field of medicine. Many drugs are too expensive for people in developing countries, so they can't get the right treatment. With clever little changes, pharmaceutical companies try to extend their patents, thus preventing other producers from marketing the same product at a more favourable price. In other words, products that have a use value cannot be made in some countries because they are protected.

In practice, this often means that the owner of the patent does not offer the products in poorer countries because people do not have the money for them.

> **CHENNAI, India, August 6 (Reuters):**
> An Indian court rejected on Monday a challenge by Novartis to Indian law that denies patents for minor improvements to known drugs, and the Swiss drug giant said it was unlikely to appeal.
>
> The closely watched case in the Madras High Court had become a key battle in the long-running war between multinational drug firms and humanitarian campaigners, who say 'big pharma' is putting patents ahead of patients. The court in the southern city of Chennai rejected the challenge, saying it had no jurisdiction on whether Indian patent laws complied with intellectual property rules set by the World Trade Organisation, as Novartis had questioned. Novartis had said a part of the law violated the Indian constitution as it was 'vague' and gave arbitrary powers to patent authorities. But the two-judge bench dismissed the challenge, saying Novartis was 'no novice' in pharmacology to not understand a law that says a patent applicant has to show that the discovery 'resulted in enhancement of known efficacy of the substance'. The objective of the Indian patent act, they said, was also to 'provide easy access to citizens to life-saving drugs'.
> *Source* Reuters, 6 August 2007[16]

Another disadvantage of patents is that they are relatively expensive to apply for. Inventors with little capital are therefore less likely to go down this path than large capital-rich companies.

2.2 Damage arising from Cost Savings

The overconsumption of (polluting) resources, resulting in excessive waste beyond the Earth's capacity, is not the sole tragedy contributing to an unsustainable world. To survive in the free market, a company is also challenged to keep costs down. Chapter 1 taught you numerous ways to achieve this cost reduction. In a study by Van Witteloostuijn (1999), cost reduction is described as an "anorexia strategy," wherein companies become excessively fixated on minimizing and controlling costs, ultimately leading to their own demise. However, this approach also gives rise to various other detrimental effects.

Damage Caused by Efficient Use of Resources

In this section, you'll learn about the harmful effects of the cost-saving measures described in Chapter 1 (economies of scale, pesticides, fraud and tax structures).

Economies of scale

As mentioned before, technological innovations have led to economies of scale and thus more efficient extraction and use of resources. But the deployment of improved technology can also be harmful. One example, about which there has been much discussion in 2019, is pulse fishing, where pulses of electricity are used to lure fish into the net. Come argue that this method is more environmentally friendly than fishing with nets, because it's more energy efficient. However, opponents argue that it is *animal-unfriendly* and especially *harmful to young fish*. In 2018, the European Parliament voted to ban pulse fishing from July 1, 2021.[17]

In livestock farming, the main objective is to increase the scale to increase profits. However, this often comes at the expense of *animal welfare*. Pigs and chickens in particular are regularly in the news. Pigs are known to be sensitive, social and intelligent animals. Yet they often only have 0.3–1 m^2 of living space. Pigs naturally exhibit a rooting behavior, yet they are frequently raised on concrete floors, depriving them of this instinctual activity. To prevent tail-biting among pigs, farmers often resort to tail docking, a practice that involves removing the tails of unanesthetized animals. Wakker Dier reports that because of the poor air quality, half the pigs have lung problems. Finally, a mother pig typically goes through an average of six reproductive cycles in just three years before being sent for slaughter, allowing her minimal time to nurse her piglets.

Chickens also suffer in battery cages and henhouses. A notable example is the "plump chicken" campaign by Wakker Dier, which shed light on the practice of rapidly fattening chickens to the point where their bodies become almost inoperable. They can no longer stand on their legs because they cannot bear the weight. Because of the cramped housing, diseases develop quickly, which are treated with antibiotics. Because roosters do not lay eggs, they are killed immediately.

Did you know that each year in Europe some 680,191,000,000 pigs and 583,142,000,000 cattle are slaughtered in slaughterhouses? That is 13 billion/ 11.2 billion per week respectively! You can find those figures on the Eurostat website (Image 2.3).

Image 2.3 A calf (*Source* Pixabay, RyanMcGuire)

Economies of scale in the transport sector have mainly led to cost reductions. Economies of scale make it possible to transport more goods at a time, yet transport movements have increased (you read about this earlier in section "Damage to Humans, Animals and the Earth"). Lower costs have led to a lower price per unit, making air travel, for example, accessible to many more people. Falling transport costs also make it cheaper to produce in countries where labour costs are also low. Combined with the use of fossil fuels, these cost savings have therefore contributed to *increased CO_2 emissions*.

Pesticides

Pesticide use has negative consequences for insects, such as flies and bees. The European Food Safety Authority (EFSA) reports that over the past 10–15 years, beekeepers are seeing a weakening of bees and even whole bee colonies disappearing, especially in Western European countries such as

France, Belgium, Switzerland, and The Netherlands.[18] In 2018, it was proven that plant protection products containing active substances from the group of neonicotinoids had a direct effect on bees. Because bees are essential for the pollination of crops, the extinction of bees poses a direct threat to our food supply (Image 2.4).

One step further up the food chain, the decline in insects has led to a *decline in meadow birds*. The use of pesticides is thus demonstrably threatening biodiversity. In addition, pesticides or plant protection products also pose *a threat to human health*. That is why the law sets a limit to the amount of pesticides that may be used on fruit and vegetables. You can find more information on this subject on the website of the European Union.[19]

Cheating

You also read in Chapter 1 about cheating with raw materials. The fraud in 2009 with baby milk powder led to babies dying and many others becoming ill. Such scandals show the *health risks* when consuming products and that consumers *can lose confidence in products*. This also applies to the many labels that companies use to show that they supply sustainable products. If you then hear later that it was just a form of *greenwashing* (presenting a company as more sustainable than it is, see also Chapter 5) and that the extra profit has

Image 2.4 Insects need nature without pesticides (*Source* Own image)

mainly ended up in the shareholders' pockets, you lose confidence and go back to buying the cheapest products.

Tax structures

The last strategy examined in Chapter 1 was the implementation of intelligent tax structures. As a result, the government of the country where the company is officially operating experiences a significant reduction in tax revenue, hindering its ability to fulfill its public responsibilities effectively. The funds are redirected to foreign governments and the company owners, thereby depriving the local government of essential financial resources. This has major consequences, in particular for developing countries, where much of the production takes place. Oxfam Novib states that 'tax avoidance by MNCs using tax havens is estimated to cost developing countries at least $100bn (approximately £78bn) every year'.[20] Tax havens thus reinforce social inequality in two ways between countries: more profit for the owners of the company and less tax revenue for the country where the work is done.

Damage Caused by Managing Employee Productivity

In Chapter 1, you learned four ways that companies encourage, entice, or coerce people to produce more for the same money (imposing targets, standardization of actions, automation and robotization, and merger of activities). These all have negative effects as well.

Imposition of targets

Imposing goals increases employee productivity. Being goal-oriented is instilled at an early age, for example, in school. On the one hand, this fits well with your capabilities: if you are good learner, you get more challenges and opportunities to develop at your level. The downside, however, is that people with less abilities receive a position in society *that many people look down on.* Cleaners, garbage collectors or people sitting at home unemployed: many people in society are quick to express an opinion about them. They should 'try harder', 'they don't want to' and 'they get money but don't do anything for it' are regular criticisms about these groups. For people with disabilities, the organization of our economy even leads to exclusion from economic life: because they can hardly or not at all produce, they are of no economic value. Government financed day care centres offer these people the possibility to spend their day in a meaningful way.

There is increasing interest in the negative effect of managing people's productivity. An important reason is that it is linked to stress and burnout at work.

> According to the World Health Organization, 'Burnout is a syndrome conceptualized as resulting from chronic workplace stress that has not been successfully managed. It is characterized by three dimensions:
> - feelings of energy depletion or exhaustion;
> - increased mental distance from one's job, or feelings of negativism or cynicism related to one's job; and
> - reduced professional efficiency'.[21]

Paul Verhaeghe, psychiatrist and social critic, argues in his book *Identity* that burnout is not the result of work pressure or working too hard, but 'the way work is organised and, in particular, the associated social relations' (2013: 219). By social relations, he refers to the hierarchical relations that prevail in companies. Because of this hierarchy, productivity is imposed. According to Verhaeghe, the less freedom you have in the organization of your work, the more ill that work becomes. A pressbutton at the assembly line is a reaction to the lack of freedom there. With that button people regained control over their work.

Speklé and Verbeeten (2014) and Ariely et al. (2009) show that setting goals, usually linked to a reward or punishment, can also lead to *poorer performance*. An important reason for this is that the human psyche is not taken into account when targets are set. Kilian Wawoe (2010) discovered during his research in the banking sector that target setting *stimulates tunnel vision or bonus blindness*. The employee focusses only on the set targets and no longer sees new possibilities and opportunities. It can also lead to *minimal commitment* from the employee regarding non-targeted activities. For example, some university employees state that they want to teach as little as possible because they are not rewarded for this in the form of a promotion to a professorship.

Research by Van der Stede (2011) shows that people only focus on *the short term* when it comes to targets. For example, if you are rewarded for higher quarterly results, you will try to keep the costs per quarter as low as possible. A long-term strategy that costs money in the short term is therefore not attractive, while it could be essential for the future of the company.

Organizational psychologists Deci, Koestner and Ryan have done extensive research on the subject of motivation. Their 1999 study showed that people perform much better if the motivation comes from within themselves. If people were driven by targets with a reward attached, this had a negative effect on their *intrinsic, autonomous motivation*. Therefore, it is better to connect to what people want themselves than to impose goals on them. You will learn more about this in Chapters 3 and 4.

Standardization of operations and work processes

Another way to increase employee productivity is standardization of operations. Karl Marx (1818–1883) called attention to the negative effects of this mode of production. Marx was convinced that work is an important source of meaning for people. With the advent of industrialization, the feasibility of individuals producing an entire product diminished. The process became time-consuming and the resulting product became significantly more expensive compared to those partially manufactured by machines. The principles of the market therefore shifted production from individual crafts to factories. In the factory, people had to perform partial operations, often on assembly lines. Even today, people still experience this type of work as monotonous and mind-numbing.

Marx's main criticism was that dividing work into small parts *alienates people from the work and possibly from themselves*. Alienation means that you no longer feel connected to what you are making. You perform the actions, but it feels empty and pointless. Your work is dictated by the machine or by the description of your work in protocols or output. Alienation encompasses the notion that individuals lose awareness of the consequences of their actions and become disinterested, primarily because they are unable to observe the direct impact of their actions. Due to the prevalence of fragmented tasks where individuals only experience a portion of the whole process, a sense of responsibility for the entirety of the outcome tends to diminish. This lack of awareness and responsibility is a pervasive aspect of daily life, whether one is consciously aware of it or not. Do you know, for example, which chicken your piece of chicken breast came from? Where your clothes are made? Have you ever pondered the origin of the trees that provided the paper for this book? Because we don't know, we don't feel guilty or responsible for the way those chickens lived, the miserable conditions in which the seamstresses had to work, or the loss of the rainforest in Brazil. Because of alienation we can close our eyes to the negative effects we are partly to blame for.

Automation and robotization

If assembly line work is boring and monotonous, then the solution of *replacing the human being* completely makes perfect sense. In addition, a fully automated process is also faster and more reliable. You don't have to deal with sick people who sometimes make mistakes, a robot never complains and doesn't suffer from boredom. The result, however, is that some people no longer have work and become unemployed. While assembly line work harms people, research has shown that work in general contributes to mental and physical health and makes people happier. For example, Sanya Singh and Yogita Aggarwal (2018) examined when people are happy at work, and found that it depends on both the individual and the organization. The most flourishing situation is when someone is self-motivated and works in an environment that facilitates them. If that same motivated employee works in a non-facilitating organization, frustration will quickly arise. Some people can be inherently less motivated as well, so if the organization does support them, this will lead to an unhealthy relationship from which the employee will benefit, but which is unfavourable for the organization. Finally, if the organization also turns its back on this employee, a destructive and festering situation arises for both parties (Fig. 2.4).

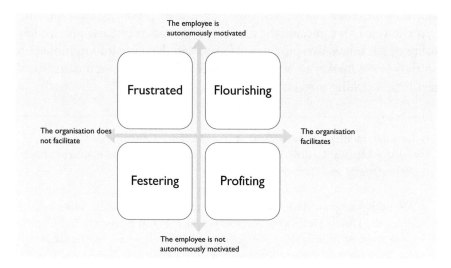

Fig. 2.4 The two dimensions for job happiness from Singh and Aggarwal

If machines or robots replace people working at jobs where those people are autonomously motivated in an organization that supports them, this will have a negative impact on them.

Merger of activities

The last way to increase productivity is by merging the same work performed by two people. If companies X and Y each have an employee responsible for purchasing toilet rolls, a merger would enable the consolidation of procurement efforts, resulting in a more efficient purchasing process for both companies. Similarly, when two small municipalities independently create policy plans for their respective areas, a merger would allow for the consolidation of these plans into a single comprehensive document. Mergers can particularly reduce the management layers and support staff. This reasoning has already led to many mergers.

Yet increased productivity often fails to materialize. Bruce Blonigen and Justin Pierce[22] examined all mergers at manufacturing companies in the United States over a ten-year period (1997–2007). They did not find reductions in costs through increased productivity and/or efficiency, but they did find higher prices and lower quality. This is because companies strengthen their market position through mergers, reducing the coercive hand of the market. With fewer companies providing similar products, customers have fewer choices. This can cause the firm to increase price, but also provide lower quality or less innovative products. Mergers can therefore lead to monopolies and thus fewer market forces. It is for this reason that governments usually prevent far-reaching mergers.

> A hospital merger between Amsterdam Medical Centre (AMC) and VU University Medical Centre Amsterdam (Vumc) in 2018 raises concerns about the affordability and quality of care:
>
> 'A healthcare giant like AMC-Vumc entails risks for public interests', wrote the Dutch Healthcare Authority, which is also present today. The regulator fears affordability and quality. Studies of hospital mergers show that they increase prices. 'Scale leads to a stronger negotiating position vis-à-vis the health insurer', explains Varkevisser. Less competition can also have a negative impact on quality: 'If you merge with your neighbour, you lose an alternative. That can lead to less incentive to work on improving quality'.[23]

Through mergers, companies grow large, and large companies have an important position in society: they provide many jobs and sometimes supply products on which many people have become dependent. If they go bankrupt, this has a major impact on society. For example, take Shell, employer of 82.000 employees by the end of 2021, with an important role in the supply of petrol and diesel. Large companies are *too big to fail*. If they threaten to go bankrupt, the government will often step in to keep them afloat. 'Too big to fail' (originally a US term) became famous in 2008, when the government bailed out banks with taxpayers' money. The problem is that if companies know they cannot go bankrupt, this leads to *unfair competition*. These large companies and banks can take many risks because they know they will eventually be taken care of by the government. With their unique position, they can also *lobby* the government for their own advantage, and sponsor future political leaders who—if they come to power—will represent their interests. Because of their size, they can free up people to lobby, and they know that the government also depends on them, for example, when it comes to employment. Lobbying is not prohibited, but it can damage democracy because the power of corporations can be stronger than the voices of the citizens.

> That smoking is bad for your health is widely known and accepted. Yet for years, the tobacco industry managed to block government measures and still has a lot of influence. How do they do it? This is what news site Forbes tells us about it: 'It turns out that tobacco major Philip Morris International spent €5.25 million to lobby Members of European Parliament, the highest for any company in the European Union in 2013... Internal documents leaked from Philip Morris' offices, which give indications of targeted efforts to delay the regulatory decision making. The company apparently used 161 lobbyists to this end'.[24]

So most mergers do not lead to economic benefits, but mergers do impact the people working at those companies. In fact, mergers are often accompanied by change within the organization, leading to *uncertainty, anxiety,* and *feelings of loss and grief*. *Stress and conflict* also arise due to cultural differences between the two former organizations. For example, the Guardian reported on July 20, 2017, that 'a clash of national cultures and an inability to understand each other's languages threatens to make the merged Air France-KLM group of airlines unmanageable'.

In addition, mergers lead to enormous companies. A much-heard voice in society is that the *human measure* is lost. A teacher knows all pupils at a school of around 300 pupils, but in schools with thousands of students children become numbers. In larger organizations, the lines between people also become more formal: in a small organization, you can solve your computer problem by dropping by the ICT officer, in larger organizations you call a helpdesk and all complaints are registered. In a small organization the management knows what happens on the work floor by regularly walking around; in larger organizations they are informed by means of standardized reports. Standardization of work processes and output therefore becomes more important the larger the organization. As mentioned earlier, this standardization can lead to alienation.

Image 2.5 We do not have to shoulder the burdens of the world alone (*Source* Pixabay, StockSnap)

We do not have to shoulder the burdens of the world alone. However, the loss of the human scale sometimes makes us feels that way. Target not achieved? On the list for dismissal at the next reorganization. Forgot to register for an exam, but were there at the right time? Too bad, you didn't follow the rules. In large organizations, it is sometimes forgotten that it is still about the vulnerable other (Image 2.5).

Damage Caused by Keeping Labour Costs Low

In addition to managing employee productivity, you also learned that labour costs can be reduced by keeping salaries as low as possible using several methods (salary scales, wage moderation, demotion, lower pay for the same job, dismissal and flexible working, offshoring and child labour). These methods also have negative effects.

Salary scales
People with the least or lowest education get the lowest pay. If you had a side job during your school years, or maybe even now, you know you have to start at the bottom of the ladder. Employers will choose to offer you the minimum (youth) wage. Have you ever thought that maybe that's a little weird? Why, for the same work, do you still get less money when you're younger? And why do people at the top receive a much higher salary? Why isn't the profit distributed fairly among all employees? Don't they all contribute to that profit?

The main reason for this inequality is that organizations are set up in a certain way; more on this in Chapter 3. People with access to capital and with entrepreneurial spirit and competences can start their own business. Profit arises because employees receive less pay than they contribute with their labour. This profit goes to the owners of the enterprise. Some companies, however, do share profits among employees, as you will read in Chapter 5.

This money can be put into a savings account, reinvested in the business or invested in company shares. There are numerous asset managers who manage the money of individuals and companies and try to get the highest possible returns through investments. And they often succeed. So good asset management can mean that if made a lot of money with your business, you don't have to work anymore. In other words: the return on capital is greater than the return on labour. Money creates money. *So a certain group of people become very rich thanks to market forces and can stop working, while others have to live on a minimum income.*

In his 2014 book *Capital in the Twenty-First Century*, Piketty showed a growing financial inequality between countries and people. Rich people are getting richer, the poor are not. A 2023 report by Oxfam International revealed that "the richest 1 percent grabbed nearly two-thirds of all new wealth worth $42 trillion created since 2020, almost twice as much money as the bottom 99 percent of the world's population" (Oxfam Novib, 2023). This *financial inequality* leads to increasing *social inequality*. Richard Wilkinson (2005) and Piketty warn that in a world where the rich get richer, and the poor do not benefit from the increased wealth, this inequality leads to *health inequalities*, *negative sentiments* and *social instability*. People wonder why they have to do mind-numbing work day in and day out for meagre pay, while they see the top of the organization driving expensive cars. How fair is that? They'd rather sit on the couch on welfare and tinker with their moped, but that's considered socially unacceptable. After all, shouldn't everyone contribute to society? But is it fair that people have to contribute to profits from which they themselves get nothing?

Social discontent was an important sentiment in the run-up to World War II, as an economic recession left many people unemployed and poor. Hitler, who had developed a strong hatred of Jews long before 1940, blamed the Jews for everything that was wrong in the world. By expelling them from their midst, everything would get better. Ultimately, his power led to a world war and the deaths of six million Jews (Image 2.6).

Image 2.6 The Jewish star, obligatory to wear on clothing by Jews in the World War II (*Source* Pixabay, JordanHoliday)

Piketty's insights were an eye-opener for many. Economists had long believed that if the rich earned more, this would automatically trickle down to people with lower incomes. Piketty's research demonstrated that this assumption often does not hold true in practice.

Financial inequality is visible in the statistics of the OECD. Topping the list is the United States, where 10% of the population owns 79% of the wealth. Simply explained: if ten people in a family have 100 euros together, one family member has 79 euros and the other nine family members may divide the remaining 21 euros: less than 2.50 euros per person. Do you understand now that this leads to resentment and dissatisfaction? The educational rule 'play together, share together' does not apply in the adult world (Fig. 2.5).

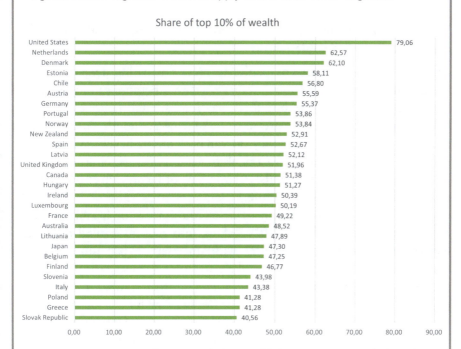

Fig. 2.5 Income inequality around the world (*Source* Own graph based on figures of the OECD[25])

This inequality also shows itself in a *difference in status*. If you are rich, people see you as successful. Rich people often have a high status, and are looked upon. Those who earn a low income or depend on social benefits often find themselves at the so-called bottom of society. This group of individuals is often subjected to negative stereotypes and unfair judgments, with assumptions being made about their work ethic or intelligence.

The classification into job scales and the status linked to this also *undervalues preparatory and intermediate vocational education*. Parents will encourage their children as much as possible to go to at least highschool. The higher the education, the lower your job scale. Many people have a negative image of vocational education, even though it trains people who are desperately needed in society (e.g., caretakers, hairdressers and plumbers).

Wage restraint, demotion, lower pay for the same job, dismissal and working with flexible labour

The flexibility needed for companies to meet fluctuating demand leads to flexible contracts and thus *insecurity for employees*. Because earning money is an essential part of life—to buy food, pay for a (rental) house and so on—many people prefer a steady stream of income every month. After all, most people don't have rich parents they can fall back on financially. With a steady stream of income you can go to a bank to get a mortgage, build savings and know that you can pay your bills. A basic flow of income gives you peace of mind and confidence for the future.

Research has been conducted to assess the impact of flexible contracts on individuals' lives (Blom, 2019). Taking into account natural differences between people (some cope better with uncertainty than others), people without a fixed contract are more likely to suffer *burnout*, arising from the pressure to keep performing because otherwise their contract might not be renewed. This group of people is also *less satisfied* with their job than people with a permanent contract. People with a temporary contract are also *less willing to make long-term decisions*, such as buying a home.

In addition, the International Monetary Fund (IMF) stated in 2018 that flexibilization is *holding back wage growth*.[26] The economic growth does not reach flex workers because clients can often offer low rates. This development increases income inequality. The OECD also indicates that competition from self-employed workers inhibits wage growth for employees with a permanent contract. The OECD, therefore, advocates government measures to strengthen the position of the self-employed in the market.[27]

However, the world has also become more uncertain for people with permanent contracts. A job for life no longer exists. Even a fixed contract can be terminated by a reorganization, you can be demoted or your salary can drop. For this group of people, it means that they can *no longer meet financial obligations* they entered into based on the expectation of a constant flow of income. People who had been doing well for years can also run into financial problems. During the recession of 2010 in particular, people could no longer pay their mortgages. In the worst case, they had even taken out an extra mortgage when the surplus value of their house was high. However, selling their house when the need arose also turned out not to be that easy, because the housing market was stuck due to the recession. As a result, this group of people ended up deeply in debt, sometimes with the result that they had to live on the streets without a job or a home.

> While there is a lack of comparable and reliable data in Europe, estimates show that there could be as many as 700,000 homeless people on any given night in the European Union. This implies that about 4.1 million people in the EU are exposed to this kind of homelessness each year.[28]
>
> If you want to know more about the state of homelessness in Europe, the website of FEANTSA (European Federation of National Organisations Working with the Homeless) is a good starting point (Image 2.7).

Image 2.7 Homeless man fighting societal prejudices (*Source* Pixabay, Island-Works)

A drop in income is often accompanied by *psychological disorders,* such as *mood swings.*[29] If employees receive a low wage for a long time while they see the company raking in profits that only the top of the organization benefits from, this can lead to *discontent* (as discussed earlier in this paragraph). This discontent manifests in staff strikes and protest votes during political elections. A protest vote is giving your vote to a politician who strongly disagrees with current government policy. With their vote, people indicate that they want a change.

Offshoring

When companies relocate their production to low-wage countries, it leads to a reduction in domestic jobs, while the products manufactured in those countries are imported back. A trade deficit occurs when a country imports more goods and services than it exports. One of the issues associated with a trade deficit is that it involves importing goods from foreign producers, which means that domestic citizens did not participate in the production process and, as a result, did not earn income from it. In countries with a trade deficit, employment often lags behind or even disappears completely, with consequences for the spending of citizens in their own country. Less spending means less production, and this spiral leads to *economic decline*. But high exports can also have a disadvantage: food producers in low-wage countries choose to export instead of sell domestically because the world market yields more. This can lead to food shortages at home, while food is destroyed in countries where the price can easily be paid.

As jobs disappear through offshoring, the supply of people on the labour market increases. This puts pressure on their negotiating position for a higher salary. Paul Krugman, Nobel Prize winner for Economics in 2008, states in his column in the *New York Times* that until the early 2000s offshoring did not have much effect on wages in the home country. However, with the rise of imports of goods from low-wage countries, there has been increasing pressure on wages in the home country (for Krugman that is the United States).[30]

The Guardian, 16 February 2017 (edited)[31]

As a call centre worker, I saw how employees are stripped of their rights

> In the call centre, workers are constantly watched. Every action is logged, from the number of sales made, to the time spent on calls and the length of breaks taken—measured to the second. Because calls were recorded, errors—which were hard to avoid under such high-pressure conditions—were used to discipline and fire workers on the spot. Adding to the weight of surveillance were gruelling targets. These targets were

> displayed on whiteboards at the end of each row of desks. A large TV hung from the ceiling showing a running total of how many sales each worker made, ranked in order.
> This environment was psychologically draining for many of the workers and created a tangible feeling of precariousness. This precariousness was used as a managerial strategy to discipline and motivate workers, creating an oppressive, stressful, and exploitative workplace. The workers had to regulate their own behaviour, knowing that at any moment they could lose their job.

A frequent argument in favour of offshoring is that it provides employment and thus income in low-wage countries. However, these large companies squeeze small local entrepreneurs out of the market. After all, these companies can offer their products at a lower price in the country of production than local, small-scale businesses can. Nowadays, the cotton in our clothing travels halfway around the world before it is available in stores. Every link in that chain needs a profit to survive. If you buy cheap jeans at Zara, you may assume that Zara makes a profit. But how big is the margin for the producers of cotton in Kazakhstan, the yarn spinners in Turkey, the yarn dyers in Taiwan, the weavers in Poland and the seamstresses in Bangladesh? Competition is fierce in the clothing industry, and the power of the big retailers is great. They provide an outlet for many small sewing workshops in low-wage countries, which used to be independent sewing workshops, but now have to work for the big retailers because of the competition. This allows these retailers to put a lot of pressure on the price and make high demands. Sometimes a retailer even gives the same order to different sewing workshops, with the condition that only the best clothes are bought. The remainder are then left with unsold clothes.

> **Example**
> **The Rana Plaza Disaster**
> On 24 April 2013, an eight-story commercial building, Rana Plaza, collapsed just outside Dhaka. The building contained five clothing factories: most of the people in the building at the time were garment workers. Over 17 days of search and rescue, 2,438 people were evacuated, more than 1,100 people died, and many more were left with life-long debilitating injuries.[32]

Child labour

Using children is a last resort to keep labour costs down. This leads to unfair prices and thus unfair competition with manufacturers who do not use children. Child labour was gradually abolished in Europe in the 2nd half of the nineteenth century. They eventually accepted that children are entitled to a childhood and education. In addition, work is harmful to their physical and mental health. But worldwide child labour still exists in abundance. UNICEF estimates the number of working children in early 2020 at around 160 million worldwide.[33]

2.3 Summary

The tragedy of the coercive hand of the market

To meet the compelling hand of the market, firms focus on product innovation, quality, customer relationships, and low cost. But these strategies can have harmful consequences.

Harmful consequences of product innovation, quality improvement and customer relationship

Companies constantly have to come up with new or better products than their competitors. This continuous flow of new products leads to a continuous flow of extraction, consumption, disposal and logistics of (polluting) raw materials and products. This linear consumption leads to the following damage:

- overexploitation/excessive ecological footprint
- damage to humans, animals and the earth (including climate change due to excessive CO_2 emissions, with associated effects such as drought, flooding and rising sea levels, oil spills, earthquakes, disappearance of rainforests and coral, reduction of biodiversity and animal testing)
- Increased waste.

Achieving innovation requires (often large) pre-investment by companies. They want to earn this money back with a return. If possible, they apply for a patent for their innovation, so that only they can market this product. This process has two adverse effects:

- Only products that are expected to make money are developed and marketed.

- Useful products cannot be made in certain countries because they are protected. The owner of the patent doesn't sell the products in those countries because locals can't afford them.

The negative effects of cost savings

The strategies used by companies to achieve cost savings have the following negative effects:

Efficient use of resources

To be achieved by	Negative consequences
Put pressure on the purchase price	• Child labour
Scale (e.g., through technological innovations)	• Animal suffering
	• Biodiversity loss
	• Increase in aircraft movements/CO_2 emissions
Use of pesticides	• Land depletion
	• Extinction of species
	• Danger to human health
Deceit	• Potential effects on human health
	• Loss of confidence in the product
Smart tax structures	• Increasing social inequality
	• Loss of tax revenue

Increasing productivity

To be achieved by	Negative externalities
Imposing targets/the performance society	• Work pressure and burnout
	• Tunnel vision
	• Short-term orientation
	• Extrinsic motivation and worse performance
Standardization of operations	• Monotonous and mind-numbing work
	• Alienation from self, feeling empty and senseless
	• Alienation from the final product
	• Alienation from the impact of your actions
Automation and robotization	• Disappearance of jobs
	• Some people can no longer find suitable work

(continued)

(continued)

To be achieved by	Negative externalities
Mergers	• Uncertainty, fear and grief • Stress and conflict • Loss of the human dimension • Unfair competition • Lobbying power

Keeping labour costs low

To be achieved by	Negative externalities
Job profiles	• Difference in financial compensation • Difference in status/social inequality between people • Undervaluing practical education • Exploitation • Dissatisfaction among people with simple job profiles
Flexible contracts	• Work pressure and burn-out • Job dissatisfaction • People are less willing to make long-term decisions • Inhibits wage growth • Uncertain ability to meet financial commitments
Offshoring	• Domestic job losses • Pressure on wages
Child labour	• Children lose out on their childhood and schooling • Harmful to mental and physical health • Leads to unfair pricing

Notes

1. Overshoot day. Accessed September 1, 2022, from www.footprintnetwork.org.
2. Climate change. Accessed September 1, 2022, from https://climate.nasa.gov/.
3. European Commission, Directorate-General for Mobility and Transport, EU transport in figures: statistical pocketbook 2021, Publications Office, 2021. Accessed September 1, 2022, from https://data.europa.eu/doi/10.2832/27610.
4. Sea level rise and arctic sea ice. Accessed September 1, 2022, from https://climate.nasa.gov/.
5. Anote's arch. https://www.imdb.com/title/tt7689934/.
6. The world's troubling new tempo of temperature records. Accessed September 1, 2022, from https://www.bloomberg.com/graphics/2022-record-high-temperature-world-maps/.

7. The Chernobyl disaster. Accessed September 1, 2022, from https://www.nationalgeographic.com/culture/article/chernobyl-disaster/.
8. 'Absolutely shocking': Niger Delta oil spills linked with infant deaths. Accessed September 21, 2022, from www.theguardian.com/global-development/2017/nov/06/niger-delta-oil-spills-linked-infant-deaths.
9. An oil spill you've never heard of. Accessed September 20, 2022, from https://edition.cnn.com/2018/10/23/us/taylor-energy-oil-largest-spill-disaster-ivan-golf-of-mexico-environment-trnd/index.html.
10. The state of world's forests. Accessed September 1, 2022, from https://www.fao.org/state-of-forests/en/.
11. The IUCN red list of threatened species. Accessed September 26, 2022, from https://www.iucnredlist.org/.
12. Waste generated by households by year and waste category. Accessed September 1, 2022, from https://ec.europa.eu/eurostat/databrowser/view/ten00110/default/table?lang=en.
13. Fashion and the SDGs: What role for the UN? Accessed September 21, 2022, from www.unece.org/fileadmin/DAM/RCM_Website/RFSD_2018_Side_event_sustainable_fashion.pdf.
14. Report: Amazon destroys large amount of new, returned goods. Accessed September 21, 2022, from https://financialpost.com/pmn/business-pmn/report-amazon-destroys-large-amount-of-new-returned-goods.
15. The Plastic Soup Foundation. Accessed September 1, 2022, from https://www.plasticsoupfoundation.org/en/.
16. UPDATE 3-Indian court rejects Novartis patent challenge Accessed September 1, 2022, from https://www.reuters.com/article/india-novartis-idUKDEL10325320070806.
17. New fisheries rules: Add a ban on electric pulse fishing, say MEPs. Accessed September 30, 2022, from https://www.europarl.europa.eu/news/en/press-room/20180112IPR91630/new-fisheries-rules-add-a-ban-on-electric-pulse-fishing-say-meps.
18. Bee health. Accessed September 21, 2022, from www.efsa.europa.eu/en/topics/topic/bee-health.
19. Pesticides. Accessed September 7, 2022, from https://food.ec.europa.eu/plants/pesticides_en.
20. Making tax vanish. Accessed September 7, 2022, from https://www.oxfam.org/en/research/making-tax-vanish.
21. Burnout an "occupational phenomenon". Accessed September 7, 2022, from https://www.who.int/news/item/28-05-2019-burn-out-an-occupational-phenomenon-international-classification-of-diseases.
22. Mergers may be profitable, but are they good for the economy? Accessed September 21, 2020, from https://hbr.org/2016/11/mergers-may-be-profitable-but-are-they-good-for-the-economy.

23. Fear of disappearance of human scale in Amsterdam 'care factory'. Accessed September 30, 2022, from https://www.trouw.nl/nieuws/vrees-voor-verdwijnen-van-menselijke-maat-in-amsterdamse-zorgfabriek~bf466837/.
24. Why did Philip Morris spend more than anyone else lobbying the E.U.? Accessed September 7, 2022, from https://www.forbes.com/sites/greatspeculations/2014/10/03/why-did-philip-morris-spend-more-than-anyone-else-lobbying-the-e-u/?sh=32b7bea52963.
25. OECD data; Wealth. Accessed September 14, 2022, from https://stats.oecd.org/Index.aspx?DataSetCode=WEALTH.
26. IMF calls for countering labour market flexibility. Consulted on September 30, 2022, from www.nu.nl/economie/5154653/imf-roept-flexibilisering-arbeidsmarkt-gaan.html.
27. Oeso: Flexibilisation of the Dutch labour market has gone too far. Accessed September 21, 2022, from www.nrc.nl/nieuws/2018/07/03/rapport-oeso-flexibilisering-van-de-nederlandse-arbeidsmarkt-is-doorgeschoten-a1608691.
28. How many people are homeless in the European Union? Accessed September 21, 2022, from https://www.feantsa.org/en/about-us/faq.
29. Negative socioeconomic changes and mental disorders: a longitudinal study. Accessed September 21, 2022, from https://jech.bmj.com/content/69/1/55.abstract.
30. Trouble with trade. Accessed September 21, 2022, from www.nytimes.com/2007/12/28/opinion/28krugman.html?ex=1356498000&en=59380e4088506422&ei=5124&partner=permalink&exprod=permalink.
31. As a call centre worker I saw how employees are stripped of their rights. Accessed September 21, 2022, from https://www.theguardian.com/careers/2017/feb/16/as-a-call-centre-worker-i-saw-how-employees-are-stripped-of-their-rights.
32. The Rana Plaza disaster. Accessed September 21, 2022, from https://www.gov.uk/government/case-studies/the-rana-plaza-disaster.
33. Child labour. Accessed September 21, 2022, from https://www.unicef.org/protection/child-labour.

Literature

Ariely, D., Gneezy, U., Loewenstein, G., & Mazar, N. (2009). Large stakes and big mistakes. *Review of Economic Studies, 76*(2), 451–470.

Blom, N. (2019). *Partner relationship quality under pressing work conditions: Longitudinal and cross-national investigations*. Ipskamp Printing.

Hardin, G. (1968). The tragedy of the commons. *Science, 162*(3859), 1243–1248.

Oxfam Novib. (2023). *Richest 1% bag nearly twice as much wealth*. Accessed June 23, 2023 from https://www.oxfam.org/en/press-releases/richest-1-bag-nearly-twice-much-wealth-rest-world-put-together-over-past-two-years

Potapov, P., Hansen, M. C., Laestadius, L., Turubanova, S., Yaroshenko, A., Thies, C., & Esipova, E. (2017). The last frontiers of wilderness: Tracking loss of intact forest landscapes from 2000 to 2013. *Science Advances, 3*(1).

Singh, S., & Aggarwal, Y. (2018). Happiness at work scale: Construction and psychometric validation of a measure using mixed method approach. *Journal of Happiness Studies: An Interdisciplinary Forum on Subjective Well-Being, 19*(5), 1439–1463.

Speklé, R. F., & Verbeeten, F. H. M. (2014). The use of performance measurement systems in the public sector: Effects on performance. *Management Accounting Research, 25*(2), 131–146.

van Bavel, B. (2018). *The invisible hand*. Promotheus.

van der Stede, W. A. (2011). Management accounting research in the wake of the crisis: Some reflections. *European Accounting Review, 20*(4), 605–623.

van Witteloostuijn, A. (1999). *The anorexia strategy*. De Arbeiderspers.

Verhaeghe, P. (2013). *Identity*. De Bezige Bij.

Wawoe, K. W. (2010). *Proactive personality: The advantages and disadvantages of an entrepreneurial disposition in the financial industry*. www.WorldCat.org

Wilkinson, R. G. (2005). *The impact of inequality: How to make sick societies healthier*. Routledge.

3

Guiding Principles in Our Current Economic Model

The preceding chapter delved into the detrimental effects caused by the free market. Can we reduce or eliminate this damage? Or is the coercive hand of the market too strong?

The market was born out of our needs. Adam Smith argued that it was an evolutionary process, not invented by anyone, but born out of mutual interest. Analogously, you can draw a parallel to language, which similarly evolved through human interaction. Or with technological developments, driven by curious people who think about how things can be made better and easier. Language can be considered as a tool, and when we use it in a manner that causes offense, we are misusing that tool. Technology is a tool; the question is whether we use it for good or for bad. Similarly, the market can be seen as our instrument for trading products and services. In its essence, the market itself is neither inherently good nor bad. The presence of a coercive hand and the consequent harmful effects stem from the choices we make within the market.

These choices are guided by the values we hold. In this book, those values are called *guiding principles*. Without you sometimes being aware of it, your behaviour is also guided by these principles. An example is the choice of whether or not to eat meat. The same applies to the choices we make in the economy. For instance, when an entrepreneur employs someone within their organization at a wage aligned with market standards, it reflects a guiding principle regarding their perception of how individuals should interact within organizational settings. When governments opt for open borders in trade,

Fig. 3.1 The economic system

it signifies a guiding principle that "free trade is beneficial." The market, in combination with these guiding principles, is the *economic system*. If the guiding principles stimulate freedom, this leads to a free market economy. If the principles are aimed at sustainability, this leads to a sustainable market economy. And if they are aimed at control and planning of production, with the state owning all businesses, this leads to a planned economy. With guiding principles (which are sometimes quite unconscious), we steer the functioning of the market (Fig. 3.1).

Numerous countries currently endorse and promote the forces of the free market. In this chapter, you will explore eight fundamental guiding principles that form the basis of this functioning. They are divided into two main categories.

Section 3.1 discusses the principles that support the operation of the market:

1. the freedom to serve one's own interest;
2. productivity through labour specialization;
3. free trade;
4. private property;
5. money drives our activities.

Section 3.2 examines the position of human beings in the free market in more detail:

1. the position of people in the organization;
2. people's needs and motivation.

As a final guiding principle, you learn about:

3. economic growth and trickle down.

3.1 Acting in Freedom for More Possession

Freedom to Serve One's Own Interests

Most countries now have free market economies. The fall of the Berlin Wall in 1989 was widely interpreted as a triumph of liberalism, emphasizing individual freedom, and the invisible hand of the market, over communism, which was rooted in collective principles and a centrally planned economy. The overarching guiding principle suggests that individual freedom and competition are essential in fostering innovation, high quality, and cost-effective products in the market. In contrast, in centrally planned economies like the Soviet Union and China, where manufacturers were obligated to provide products at predetermined prices, the motivation for investing in innovation and quality was diminished. Furthermore, the freedom for entrepreneurs to explore alternative endeavors was limited in such systems.

The belief in the beneficial effects of the market where people trade freely with each other and with a government that interferes as little as possible is based on Adam Smith's book *An Inquiry into the Nature and Causes of the Wealth of Nations* (1784).[1] Because of his work, he is considered a founder of current economic thinking. An important question Smith wanted to answer was how to increase the wealth and power of nations. During his era, poverty was considerably more widespread than it is in the present day, and motivated by empathy, Adam Smith aimed to alleviate this poverty.

Smith examined the workings of the market in detail and developed the insight that the unrestricted actions of individuals were the driving force behind the establishment of the free market. *Self-interest* is a key concept in Smith's work. Smith reasons that if everyone acts on the basis of self-interest, this will ultimately lead to a better life for everyone. Promoting one's own self-interest in freedom acts like an *invisible hand* with which the individual contributes to this societal prosperity, without having thought of it beforehand.

So what happens, Smith thought, if that self-interest is given free rein and transactions are seen as something that serves both parties instead of, as was thought in the eighteenth century, a *zero sum game*? A zero sum game is a situation in which one party gains exactly what the other party loses. Smith argued that the economy actually benefits everyone: more can always be produced, and through exchange everyone wins. The most important source of prosperity is not what you own, but what you produce and can exchange. So not possession, but your labour power is your source to improve, he writes in book one of *The Wealth of Nations*. The more productive the inhabitants of a country are, the more they can exchange on the market, and the better *supplied* they become:

The annual labour of every nation is the fund which originally supplies it with all the necessaries and conveniences of life which it annually consumes, and which consist always either in the immediate produce of that labour, or in what is purchased with that produce from other nations. According therefore as this produce, or what is purchased with it, bears a greater or smaller proportion to the number of those who are to consume it, the nation will be better or worse supplied with all the necessaries and conveniences for which it has occasion.

Labour Specialization for Higher Productivity

Smith also taught that this higher productivity could be achieved through the specialization of labour, a phenomenon that was on the rise in his day due to emerging industrialization. Smith cites the example of the pin manufacturer, who, with machines and the distribution of labour among workers, could make many more pins than one person alone could. It also allows unskilled workers to participate in the production process: one action is quickly learned, and because they perform it continuously, they quickly become experts at that action (Image 3.1).

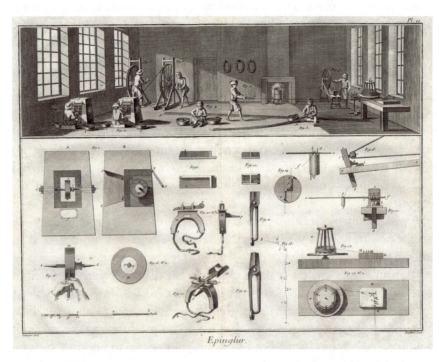

Image 3.1 A poster of a pin factory in the time of Adam Smith. The image originally appears in Diderot's Encyclopedia (1751–1772)

The concept of labour specialization has become so deeply ingrained in our work practices that it is now unimaginable to envision a society without it. This goes beyond just performing partial actions on an assembly line. Without you perhaps realizing it, all education is geared towards learning this kind of job specialization. Nurse, doctor, judge, builder, hairdresser, communication advisor, administrator or personnel officer are parts of the many facets of (economic) life. Because we specialize, we have something to trade. You cut my hair and I'll give you a piece of meat. And because you become skilled, you often become better than someone who would have to do everything themselves. Quality and productivity are therefore related to labour specialization.

A Larger Market Through Free Trade

If a company can produce a lot, it can also sell a lot. Locally, the market may soon become too small and so companies expand their sales to other cities and countries. They can also have their products made abroad and then ship them to all other parts of the world. Countries choose whether to tax sales of goods from abroad. When goods can move without import or export restrictions or subsidies, this is called *free trade*.

Free trade originated with David Ricardo (1772–1823), among others, who theorized that each country has its own comparative advantage and that countries should focus on this for economic growth.

Ricardo's theory has long been followed by many policymakers. The focus of governments after World War II was mainly on stimulating a free global market. They concluded that an economic union could be a strong means to prevent wars. After all: if you are economically dependent on each other, then a war will harm yourself. This led to the establishment of various free trade zones per continent, such as the European Union (EU), the US-Mexico-Canada Agreement (USMCA), the Asian Free Trade Area (AFTA) and the African Free Trade Zone (AFTZ). In 2019, a free trade area came into force between the European Union and Japan after years of negotiations. Globally, 164 countries cooperate in the World Trade Organization (WTO).

All free trade areas encourage the free movement of goods and capital by removing protectionist measures, such as import tariffs (a tax on what you import from another country) and import quotas (a cap on the amount that can be imported). But they also create other enabling policies. In Chapter 1 we read that the European Union's competition policy encourages companies to offer consumers goods and services on the most favourable terms, promoting efficiency, innovation and low prices.

Right of Ownership Aimed at Private Property

In order to trade something, you have to own something. People offer their property on the market, and through a transaction, the property passes from the seller to the buyer. Property is regulated by law: 'In law the term property refers to the complex of jural relationships between and among persons with respect to things. The things may be tangible, such as land or goods, or intangible, such as stocks and bonds, a patent, or a copyright' (Encyclopaedia Britannica).

Private ownership works well in a market economy because it is consistent with the principle of self-interest. By doing business you increase your property and thus your wealth. The principle encourages people to work because the result of this work serves their own interest. In addition, people are often more careful regarding things that belong to them than things that everyone owns or can use.

Companies, in essence, are a form of property, wherein individuals, shareholders, or members can become owners of a company. These enterprises are called *private organizations*. Being the owner of an enterprise means that you have the most comprehensive right to the movable (such as means of production, raw materials and the like) and immovable property acquired with this enterprise. You do not own the people who work in the company, but you do own the employment contracts. Forms in which the owner is liable with their private assets for these goods, and for any debts of the company, are called unincorporated legal forms, for example, sole proprietorship, or partnership. An entrepreneur can also make the company the owner of (and therefore liable for) the business. In that case, the company is the legal entity and not the entrepreneur. The entrepreneur is then often the director of that legal entity. Well-known examples of this are private and public limited companies. As a private person, you run less risk, because the liability stops at the company assets.[2]

The crucial connection between financing an organization and exerting control over its property is vital. In legal forms without corporate rights, individuals themselves invest the money, whether borrowed or otherwise, and retain decision-making authority over its usage. In legal forms with corporate rights, such as private or public limited companies, shareholders hold significant power. Ownership of shares grants voting rights during the general meeting of shareholders (ava). However, to ensure the practical management of a company, this power is often delegated to an elected board of directors, tasked with representing the shareholders' interests. Financing methods that do not entail control include bank loans or crowdfunding platforms.

Against the private organization stands the *public organization*. Rather than a private owner, public organizations are owned by the government. In a free market, the government chooses to own as few organizations as possible. For example, confidence in the benefits of market forces led to the privatization of many government organizations in the 1980s and 90s (the government sold the companies it owned to shareholders). Examples are privatization of telephone lines, mail delivery, railway maintenance and energy supply. In The Netherlands alone, the government owned 1105 companies in 2007 (first quarter), by early 2019 this had more than halved to 505 organizations (CBS figures).

> Norway has an example of a well-run state enterprise that offers benefits to all the country's inhabitants: its oil extraction company, Statoil. The state invests the resulting revenue in its oil fund. This capital makes Norway the richest country in the world. The interest pays for numerous collective provisions. Equal wealth for all, instead of just for the owners of a private company.

In some countries, most (or even all) companies are owned by the government. Well-known examples of this used to be the Soviet Union and China, but even these countries now have the protection of private property. Countries that are still truly communist are Cuba, Laos, North Korea and Vietnam.[3]

Monetary Value Drives Our Activities

"What holds value?" is a profound and philosophical inquiry that elicits a range of perspectives and responses. Values encompass diverse aspects that individuals consider important. They span from essential necessities such as clean air and drinking water to cherished experiences like visiting loved ones, from material possessions such as a well-maintained car to practical considerations like warm attire in cold climates, and from intangible bonds such as love to various personal principles and beliefs. Some like to spend their time reading a good book, while others prefer to write a book because they have something to tell the world. Considering that our time on Earth is limited, it is common for people to desire to make their lives as meaningful and fulfilling as they can.

But we also have to earn a living. That implies that it is important to provide something that others find valuable. For instance, if you consider picking up cigarette butts every day valuable (as it contributes to a cleaner environment), but no one is willing to compensate you for it, you may need to make alternative decisions about how you allocate your time (unless you are already financially secure). If you are very happy making art, but no one appreciates your works of art, you might have to find a job on the side to earn money.

Your product or service must have *monetary or exchange value* on the market. Those can differ from *use value*. Examples of products with use value but without monetary value are oxygen, sunlight and nature. Water could also belong here, but we still pay for it because of the purification costs. Sunlight has a use value because we can relax in a nice sun, and a monetary value through solar panels.

The appreciation of others for your work and the question of whether the other person is willing to pay for it is therefore an important guide in the choices of people and companies. Money largely steers our activities. Once there is a basic income or a company makes enough profit, only then will there be room for activities that do not generate money.

Earning money also controls the extent to which costs are incurred for that activity. For example, suppose a machine builder can make two machines: machine X can be made more cheaply than machine Y, but has higher CO_2 emissions. As long as the customer prefers a cheaper machine with higher CO_2 emissions to a more expensive one with lower emissions, the machine builder will continue to supply machine X. After all, they too have to make money. Another good example is the question of how many costs a company is prepared to incur to guarantee safety. For example, should all planes be grounded to solve a software problem, or is that too expensive and is it therefore more lucrative to risk a certain number of crashes per year? You may think this is an exaggeration, but read the example in the box below.

Michael Sandel, professor of political philosophy at Harvard University, in his third lecture of the online course *On Justice*,[4] gives a famous example in which a choice had to be made between less profit and human lives. It is the story of car manufacturer Ford. Ford employees discovered in the 1970s that the gas tank of their car Ford Pinto could catch fire. They drew up a cost–benefit analysis, comparing one human life against repairing all cars. Table 3.1 shows their calculation.

Table 3.1 A trade-off between costs and benefits

Cost Numbers	Value ($)	Total	Cost Numbers	Value ($)	Total
12.5 million car repairs	11	137 million	180 deceased	200,000	36 million
			180 injured	67,000	12.06 million
			2000 cars replaced	700	1.4 million
Total cost		137 million	**Total cost**		49.5 million

So when it comes to valuing the products and services of others, the consumer plays a crucial role. Ford decided not to recall the cars based on this math. It's similar to a much more recent story from 2019 from the Boeing 737 Max. Boeing discovered well before two planes crashed that the software in the plane was flawed. The faulty software caused the nose of the plane to be pushed down when it was supposed to rise. Only after two crashes did the manufacturer ground the Boeing 737 Max to solve the problem. Was a similar cost–benefit analysis performed at Boeing?

So when it comes to valuing the products and services offered by others, the consumer plays a crucial role. As consumers, we make decisions in which we consciously or unconsciously weigh and prioritize different values against each other. Do we choose organic products in the supermarket? If you choose a cheap pork steak, you are also indirectly making a decision about the welfare and life conditions of the pig from which it came. Do we choose to purchase clothes that may be slightly more expensive but are not made by children? If we opt for cheaper options, we are indirectly supporting child labor. The reason why this connection is often not strongly felt is because it is far removed from our immediate sight and there is no emotional connection with the makers of the products.

The distinction between use value and exchange value creates a gap between what is important for individuals and what is important for society as a whole. In order to secure our personal survival and maintain a certain level of prosperity, we focus on activities that generate income by offering products and services that are in demand in the market. However, if we aim to ensure collective survival and strive for a sustainable world, we must also prioritize endeavors that contribute to other forms of value, such as clean air or CO_2 reduction. The challenge arises when these additional investments in clean air, sustainable practices, waste reduction, safety improvements, and other value-driven initiatives do not yield immediate economic returns. In such cases, they are often neglected in favor of profit-driven models that solely focus on the financial gains from the products and services provided. The market is primarily driven by profit-oriented considerations, which can hinder the adoption of practices that promote broader societal and environmental well-being.

3.2 The Employee Is a Means of Production

The role of people in the market remains to be discussed. After all, it is people who produce in this system—people who are free, own property, practice their profession, trade globally and receive and spend money. People transform nature into something valuable that other people are willing to pay for. Our capacity to work is the basis for our survival. In essence, we are all a means of production. Or are we more than that?

Humans have been working for as long as they have existed. Formerly as hunters and food gatherers in small communities, in which everyone contributed to gathering and preparing food and other activities necessary for survival. Everyone from a certain age had a task, and the relationship between people was *reciprocal*: if you do something for me, I do something for you. When humans began to settle in one place, there was room for agriculture. Successful agriculture provided surplus production and supplies and thus allowed people to spend time on other things. Labour specialization came about and the exchange of goods created the market. Lucassen (2013) indicates that this labour specialization through agriculture lasted for centuries until the first cities arose around 7000 years ago. The reciprocal relationship between people changed along with it. Eventually, four labour relations between people emerged:

- *Feudalism*: An economic system in which a lord made his land available to people (serfs) to grow their food, but for which a part of the proceeds had to be given to the landlord. Serfs were also obliged to fight to protect the land.
- *Slavery*: Where people were captured and forced to work for their owners.
- *Wage labour*: where a person offers their services in exchange for wages.
- *Self-employment*: Again, someone offers their services, but they are not employed by another person. It differs from the former reciprocal relationship in that the person receiving the service does not perform a direct service in return, but pays the self-employed person.

As of 2019, wage labour and self-employment were the main labour relations in most countries.

Employer–Employee: Who Pays, decides

Working as an employee or being self-employed is considered the norm by many individuals. You could say that it is an unconscious guiding principle of the market economy. You offer your services and get money in return. In Chapter 2, you read about many harmful social effects, such as burnout or alienation. How working for someone else is related to these harmful effects is explained in this section.

At school you acquire many skills. With those skills you can start your own business to offer your products on the market. Everyone who can contributes to the market. However, most people do not start up their own businesses but sell their labour. This transaction is recorded in an employment contract. These contracts are part of the movable property of the company.

But the crucial aspect is that through this employment or contractual relationship, the employer or client also obtains control over the activities of the employee or self-employed person. Working for someone who pays you means giving up your autonomy. Those who pay the piper call the tune. This means that someone gains power over which activities you have to perform, and therefore also has the power to decide on all the instruments discussed in Chapter 1: how productive people have to be, how much they are going to earn, and whether they get a temporary or a permanent contract. Through the employment contract, the position of the employee becomes subordinate to that of the employer. Another name for an employee is therefore a subordinate. This type of organization is called a hierarchical structure. The employment contract regulates the order of precedence between people.

> **Tip**
> Unequal position between employer and employee in organizations is a recurring theme. Peter Block (1993) in his book *Stewardship* writes: '*The traditional role of line management is to be in charge of patriarchy, their primitive statement to employees being "we own you". To balance this, human resources has been put in charge of paternalism. Their primitive statement to employees is, "don't worry so much about the fact that they own you, because we will take care of you". This combination creates the golden handcuffs that make living in a world of dominance and dependency so tolerable*' (p. 147) (Image 3.2).

Image 3.2 Hierarchy in organizations (*Source* Pixabay, geralt)

This hierarchical construction is a guiding principle widely used in the market economy and has given rise to the word capitalism, in which the employer is called the capitalist. These words are based on the work of an important critic of capitalism: Karl Marx.[5] In *Capital*, he stated that wage labour (fixed and flexible) leads to financial and thus social inequality.

The core of his critique, summarized by Richard Wolff and others in multiple books, videos and podcasts,[6] is that there is a dichotomy between people: employer and employee. This dichotomy has long existed, including in the other forms discussed in the section above. In slavery there were masters and slaves, in feudalism there were landowners and serfs. In all three economic systems, there is a common element: there is *a majority that performs labour to produce a so-called surplus (you produce more than you need) for a minority that makes decisions about what happens to that surplus*. The essential difference with slavery is that you are a free person: you can choose where and for whom you want to work, or you can become an entrepreneur and thus a boss. Also, your place of birth no longer determines what you will become: landlord or feudal man. You are now free to become who you are, but you will have to earn money to support yourself. If you have no entrepreneurial qualities, you will have to offer your services to employers. Welfare and other benefits are only for people who cannot work.

The surplus (profit) in most organizations is not shared with the employees or self-employed workers, even though they have contributed to it with their labour. The profit goes to the top of the organization: the bosses and shareholders. Money therefore creates money: profits can be reinvested in one's own company or in other companies. This leads to capital accumulation in a small group of people, while many others worry every month about how to pay their energy bills or for their children's studies. This financial inequality leads to increasing social inequality, leading to the consequences described in Chapter 2.

Why did we, not all remain independent craftspeople? How has the shift to hierarchical employer-employee relationships occurred? An important factor in the shifting labour relations is the rise of industrialization. With industrialization, the craftspeople lost their independent position because they could not compete with large factories. Farmers were also driven from their land, at the time an important source of income and food supply. These people were forced to offer their skills to the owner of the factory.[7] The manufacturer paid many people the minimum wage, which in Marx's time was set at the minimum necessary to live. It resulted at that time (eighteenth and nineteenth centuries) in miserable conditions and high death rates among the paupers of the population. Marx writes extensively about this exploitation. To illustrate, the following is an excerpt about the condition of workers, including their children, in the pottery industry around 1860:

> Mr. Charles Pearson, until recently a surgeon at the same hospital, wrote in a letter to the committee member Longe, among other things: "I can only speak from personal experience and not according to statistical data; but I have no hesitation in stating that again and again my indignation was aroused at the sight of these poor children, whose health was sacrificed to satisfy the greed of their parents and employers." He enumerates the causes of the diseases of the workers in the pottery industry and he concludes this enumeration with long hours. The committee's report expresses the hope that "an industry, which

occupies so important a place in the world, will no longer bear the blot of the fact that its great success is accompanied by physical degeneration, very extensive physical suffering and the early death of the working population, which by its labour and skill has been able to achieve such great results." (Marx, 1867: 190)

Next to this, the industrialization had another effect, as described in Chapter 2, namely that technology began to determine the pace and work of man. As a result, people became subservient not only to the owners of enterprises but also to the influence of technology:

I used to carry this boy of mine on my back through the snow when he was 7 years old, and he usually worked 16 hours. (...) Often I got down on my knees with him to feed him while he was at the machine; for he was not allowed to leave the machine or stop it. (Marx, 1867: 191)

The role of the trade union and the works councils

Trade unions emerged as a response to the imbalance of power between employers and employees. Through collective action and solidarity, workers united to challenge the authority of company owners and advocate for their rights and interests. 'Trade unionism (also called organized labour) originated in the 19th century in Great Britain, continental Europe, and the United States'.[8] In the twentieth century, work councils were added within organizations, as a local representation of trade unions, but functioning independently. Trade unions and works councils have achieved much for the position and conditions of employees. Some examples are a 6-day (and later 5-day) work week with a maximum of 8 hours per day, and social benefits, labour protection and old age pension. Many countries still celebrate Labour Day, which symbolizes 'the fight for decent working conditions, job security and a decent income. The labour movement led efforts to stop child labour, give health benefits and provide aid to workers who were injured or retired'.[9]

However, co-determination was characterized as a struggle then and continues to be seen as a necessary evil in many organizations today, because the union is an addition to the existing hierarchical structure in organizations. According to Professor Richard Wolff, is akin to improving conditions for slaves, without violating the fundamental principle that it is wrong for one human being to be the property of another.[10]

Marx's work is still topical, and that awareness is growing. However, Richard Wolff argues that—especially in Western countries—his work has long been ignored because his name is also associated with communism. Marx wrote the *Communist Manifesto* (1848) much earlier than he wrote *Capital*. In it he argued that private property should be abolished, and that factories and means of production should be in the hands of the state (see the list of recommendations in the manifesto at the end of Chapter 2). However, this essentially changed nothing in the unequal relationship between employer and employee, ownership had merely been displaced. In the past, communism was at odds with the idea of free man. Rutger Bregman states that *'because of the Soviet Union, the word 'communism' became forever tainted with dictatorship and mass murder. Nobody of my generation, born around or after the fall of the Wall, would ever dare call themselves "communist" again'*.[11] The concern is that the state seeks excessive influence in the personal lives of its citizens, dictating how things should be done and limiting freedom of expression and personal autonomy. These fears, according to Richard Wolff, are all connected with the word communism.

People Need to Be Motivated to Be Productive

Before industrialization, craftsmen and farmers worked diligently as their efforts directly impacted their income. They could only produce as much as they were capable of in a day. However, with the onset of industrialization, their work shifted from individual homes and workshops to large-scale factories. Here, their labor became integrated into a broader production process, often involving repetitive tasks dictated by machines. They earned less than their added value; the profit went to the top of the organization. They had little or no say in the matter. What do you think such a shift from self-employed to factory worker does to people's work ethic and attitude towards work? Looking through the lens of the free market, the answer is that the worker wants to do as little as possible.

The foundation of the market is self-interest. Putting on capitalist glasses, the self-interest of the owner of the business is clear: produce as much and as cheaply as possible and sell it at the highest possible price. The profit is for the owner of the organization. But what is the employee's interest? The agency theory has had a lot of influence on the way organizations are run and thus the way employees are managed. This theory describes the relationship between a principal (the owner of the company) and an agent (the person who performs the work by order of the principal). An important assumption in this theory is that the employee's attitude is *opportunistic and lazy* (Corbey, 2010). Opportunistic means that the employee 'is not guided by principles, but uses circumstances to his or her own advantage' (VanDale definition). Seeking advantage for oneself is consistent with free market theory. Doing as little as possible within hierarchical positions pays off if you get a fixed wage and are protected against dismissal. You find people with this attitude (but also the opposite attitude) in every organization. Georges and Romme (1999) say that hierarchical relationships have a negative effect on people's cooperative behaviour and their productivity.

To combat opportunism and laziness, many companies have introduced variable pay and bonuses linked to the achievement of goals (*targets*). These are ways of *motivating employees in a controlled way*. This term comes from the self-determination theory of Edward Deci, Marylène Gagné, Richard Ryan and others. Controlled motivation means that an organization creates external stimuli that make people want to achieve a certain goal. You study for a test because it leads to better grades and, ultimately, a diploma. You strive to deliver high-quality work because it earns you recognition from your supervisor. You aim to sell more for the company because it translates into a higher year-end bonus. Do you know the expression 'holding out a carrot to someone'? This type of motivation works from a market perspective: by committing people to a share of the value they produce, the company's interest also becomes their own interest (Image 3.3).

Companies can also focus on increasing employees' *autonomous motivation*. Autonomous motivation means that the will to achieve a goal comes from within yourself, and you do not need external stimuli to be motivated (Gagné & Deci, 2005). Companies can increase autonomous motivation if they address the three basic psychological needs of every human being: *autonomy, competence and connectedness*. Autonomy is the space a person has to organize their work and make decisions independently. Competence means the ability to get the job done and to deliver quality. Finally, connectedness refers to collaboration and the desire to experience that our efforts are recognized by others and that we are part of a greater whole. Research shows that autonomous motivation predicts greater job satisfaction, engagement and performance, as well as less burnout than controlled motivation (Deci et al., 2017; Rigby & Ryan, 2018; Van den Broeck et al., 2009).

With theories of motivation, companies possess a powerful tool to inspire and encourage employees to perform their best, thereby contributing to the organization's profitability. But this still occurs in the context of the subordinate, hierarchical working relationship. Menno de Bree writes a critical blog

3 Guiding Principles in Our Current Economic Model

Image 3.3 Controlled motivation means that an organization devises external incentives that make people want to achieve a particular goal (*Source* Pixabay, mohamed_hassan)

about this on the website of The School of life.[12] He argues that we largely look for happiness in our lives in two fields: love and work. Through work, we can develop ourselves, grow personally and realize our authentic selves. This gives meaning to work and to life. De Bree argues that companies make clever use of this ideal: 'the enormous need for the possibility of authentic self-creation is the key to self-exploitation that capitalism has been looking for'. We work hard for the boss because we believe that hard work leads to self-actualization and happiness. Career-making is how we demonstrate personal growth. By believing in this, the problem of opportunism and laziness is solved: working hard has become in our own interest. Look how we develop ourselves! Controlled motivation is no longer necessary; because we want to work hard ourselves (be autonomously motivated), we attach less value to bonuses, salary increases or co-ownership.

> **Reflection**
> The content of Sect. 3.2 invites us to ponder the following: have we collectively embraced a fairy tale-like belief in the correlation between self-development through hard work and greater happiness? And if so, how ethically justifiable is it to rely on autonomous motivation within a market system where only the owners of the company ultimately reap financial benefits from the toil and efforts of others? Or perhaps, are we posing the wrong questions altogether? Should we simply accept work as a necessity, while recognizing that autonomous motivation is a welcome byproduct when engaged in tasks we genuinely desire?

3.3 Economic Growth and *Trickle Down*

The seven guiding principles discussed in the previous two sections are essential elements of a market economy. They form the basis for economic development and growth. This economic growth is not only an outcome but also a goal of market forces. Therefore, economic growth is the eighth and final guiding principle (Fig. 3.2).

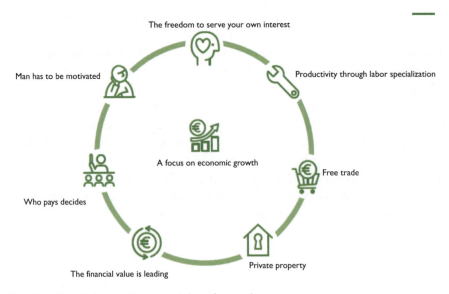

Fig. 3.2 The eight guiding principles of a market economy

Economic growth entails an increase in production, resulting in higher per capita earnings. Production, in this context, refers to the creation of added value. To provide a concrete example, let's say you engage in photography, have the photographs printed into cards, and then sell them. The added value in this process can be measured by the difference between your earnings from selling the cards and the printing costs incurred. Your labor contributes to enhancing the value of the paper and photos, enabling you to generate income that supports your livelihood. When you machine-clean cotton from the plantation and sell it to the garment industry, your machines have added value to the cotton. All the value added by entrepreneurs in a country makes up the Gross Domestic Product (GDP) of that country. This is the measure of economic growth and prosperity.

In essence, GDP is a measure of the extent to which people can generate income from their activities and make a profit, whether through labour or technology. It says nothing about the activities that people perform for which they do not receive money. Examples are raising children, voluntary work and informal care. If a country receives more income now than in the previous period, we call it *economic growth*. If a country receives less income now than in the previous period, we call it *economic contraction*. Economic growth is an *outcome* of all our economic activities (activities we can exchange for money). Economic growth worldwide has dramatically improved people's physical quality of life. But this prosperity has not been shared equally between people and across countries. You'll read more about that later.

> The late Swedish statistician Hans Rosling made it his business to show how well-off people are in many countries, contrary to what you might think if you judge by what you often see in the news. He opened many eyes with his TEDTalk in 2006.[13] His son Ola Rosling continued his work on the website https://www.gapminder.org/. On this website you can play with official data from, among others, the United Nations and the World Bank. With his work, he made more people aware of important facts about the state of the world.
>
> For example, our life expectancy has increased, related, among other things, to the level of income per person (Fig. 3.3).

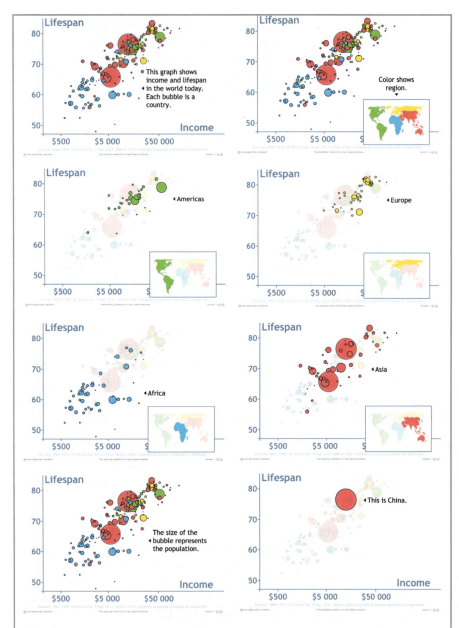

Fig. 3.3 A graph by Hans Rosling, relationship between income and life expectancy (*Source* Based on a free chart from www.gapminder.org/teach)

3 Guiding Principles in Our Current Economic Model 87

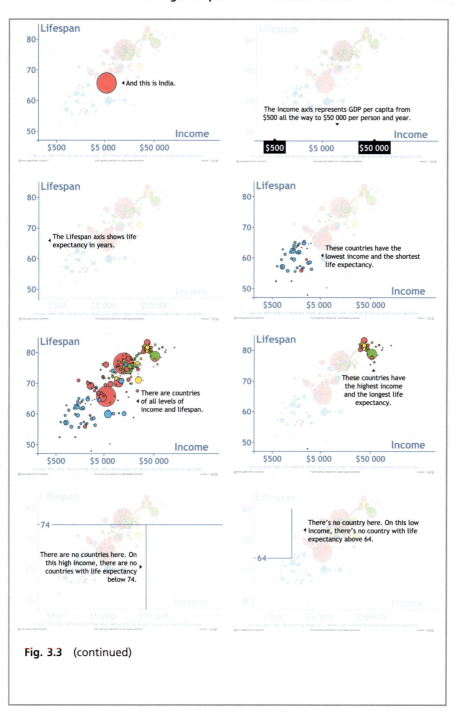

Fig. 3.3 (continued)

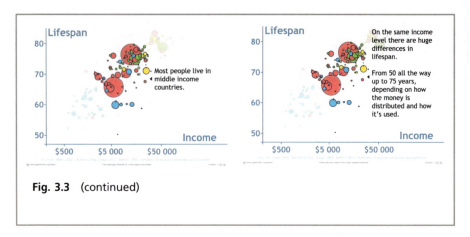

Fig. 3.3 (continued)

The number of children (0–5 years) dying has decreased dramatically, from 436 per 1000 children born in 1800 to 48 children per 1000 in 2012, globally. Again, there is a relationship between per capita income, improved health care and education. Rosling also shows in his Ted talk that the world population is growing rapidly to about 11 billion by 2100, but that in the second half of this twenty-first century, the rapid growth will slow down. This is because more women are using contraception, reducing the number of children per family. Other effects scientists attribute to economic growth include better food, fewer diseases, and a shorter work week.

The correlation between economic growth and a decrease in poverty (along with its related impacts, as mentioned earlier) has been substantiated by scientific research conducted by the OECD. When our collective income increases, we have the capacity to allocate a portion of that wealth, through taxation, towards enhancing healthcare, providing assistance to individuals in need, and establishing a basic income for those unable to work. Companies and the government can also invest in innovations that make life better and more pleasant. In addition, there is a causal relationship between personal income and health (Wilkinson, 2005).

> The central lesson from the past 50 years of development research and policy is that economic growth is the most effective way to pull people out of poverty and deliver on their wider objectives for a better life (OECD).[14]

This causal connection has given rise to the concept of *trickle down*. It posits that as economic growth occurs, the increased wealth generated at the top of society will gradually trickle down to individuals with lower incomes.

Unlimited economic growth and economic crisis

So economic growth is an outcome of all our economic activities. In current economic thinking, this growth is unlimited. Raw materials are inexhaustible, the emission of harmful gases and CO_2 is invisible, as is the pollution of the sea and air by plastic. If economic growth stagnates, we will have to consume more. More stuff, more consumption of raw materials, more pollution. Is that what we want? But what if we stop producing more economic value? In countries where we have already risen far above the poverty level, should we still focus on economic growth (with the same population)? Or is stabilization also an option?

Hans Stegeman, an economist at Triodos Bank, also poses these questions.[15] He references Benjamin Friedman, who, in his book *The Moral Consequences of Economic Growth*, asserts that sustained economic growth is necessary to uphold social stability. The prevailing belief is that individuals expect their incomes to increase and their overall well-being to improve over time. When such improvements do not materialize, or when people experience a decline in their economic circumstances, it can give rise to discontentment. If dissatisfaction reaches a critical point, it can lead to social unrest and intolerance, creating an environment conducive to the emergence of politicians who scapegoat certain population groups for economic decline. The Second World War and the economic recession that preceded it are still a lesson to many.

Economic growth and economic recession are intertwined in a market economy. Periodic economic crises often give rise to recessions, characterized by a significant contraction in economic activity. During such crises, people tend to reduce their spending for various reasons, leading to a decrease in consumer demand. This reduction in demand creates challenges for companies as they face a surplus of products and excess means of production. To adapt to the changing economic conditions, companies often need to implement cost-cutting measures to restore balance in their operations. In Chapter 1, it was discussed how companies can implement cost-cutting measures, which can involve laying off employees. As a result, individuals who lose their jobs experience a reduction in income, leading to financial constraints and difficulties in meeting their financial obligations such as mortgage or rent payments. The cycle continues as reduced consumer spending further impacts businesses, potentially leading to additional layoffs and a downward spiral in the economy (Fig. 3.4).

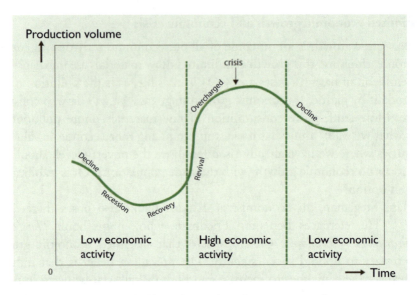

Fig. 3.4 In an economy, growth and recession alternate every few years

> An economic crisis with great impact was the 2008 crisis, which started with a financial crisis in the United States. De Nederlandsche Bank describes how this crisis started[16]: 'In the United States many so-called subprime mortgages were sold at the beginning of this century. These are mortgages for people with a low income. The interest rate on such a mortgage is usually variable and initially low, as a lure. People were told that after a few years (when they would have to pay a higher interest rate) they could take out a second mortgage on their house. That wouldn't be a problem because home prices would go up anyway. That way, many poor Americans could still buy a house. Interest rates in the United States went up across the board in 2007. Interest rates on new (second) mortgages rose as well. At the same time house prices collapsed. Obtaining a second mortgage became more difficult and more expensive than expected. Many poor Americans ran into financial problems and could no longer pay their mortgages. Defaulters were evicted from their homes and the house was sold by the bank. But because house prices were low, the sale yielded little. There, in 2007, was the trigger for the start of the credit crisis'. This credit crisis then turned into a global economic crisis.

3.4 The Role of the Government in the Market Economy

A free market is a market where people in principle are free to do whatever they think is right. However, you have already learned that in many countries the government plays an important role in making that market work. In this section you'll learn what that role is.

Government represents the collective population and serves as a mechanism for regulating collective affairs. Through the act of voting, we express our preferences and indicate the policies we deem important for both the immediate future and the long term. This includes our vision for economic policy. The economic debate is often about the question of the extent to which the government should and may interfere in the market. Should the government stimulate free trade? Where should the government restrict individual freedom? Which activities should be performed by public organizations? And to what extent does the government stimulate or restrict a flexible labour market? These types of questions can be placed in a spectrum (see Fig. 3.5):

There are three main directions an economy can take:

1. *Fully free market operation*: the government refrains from intervening and allows suppliers and buyers to freely conduct their economic activities. This approach, often associated with *liberalism*, emphasizes maximizing individual freedom by minimizing government regulations and restrictions.
2. *Mixed economy*: The government intervenes, albeit to a limited extent, in market forces in two main ways:
   a. *Creating conditions* to ensure the smooth operation of market forces. This is *a fundamental principle of neoliberal policy*.[17] Especially in the 1980s and 1990s, there was a growing belief among policymakers that a neoliberal policy would solve a number of problems. The promotion of competition and the proper functioning of the invisible hand

Fig. 3.5 The spectrum of the market on which government operates

is central to neoliberal policies. The Competition Act, based on European competition law, serves as a notable example of a governmental instrument aimed at overseeing and promoting the free functioning of the market. This legislation prohibits the establishment of cartels, which involve cooperative actions by companies to undermine competition, such as reaching price agreements. As a consequence, consumers end up paying higher prices compared to a scenario where both parties compete with each other. The European Commission, therefore, monitors the operation of the market and imposes fines. It tests whether a proposed merger of companies leads to a monopoly. The European Commission also examines whether unauthorized use has been made of state aid in the form of subsidies. After all, subsidies can potentially result in unfair competition by creating an advantage for entrepreneurs who have received them, compared to those who haven't other policies that create conditions are patent rights and encouraging cooperation between knowledge institutions and entrepreneurs.

b. The second way the government intervenes in the market is by limiting the negative externalities of market forces. In this role, the government has a duty to protect its citizens and provides a social safety net for people who cannot participate in the market. There are many examples of this, ranging from requirements for food safety, social assistance and the Unemployment Act, and agreements to reduce CO_2. Other examples of government regulation are the tax on tobacco and the ban on the sale of hard drugs. In a mixed economy, the government also owns, pays and regulates a number of organizations that mainly fulfil a public (social) function. Examples are the police, education, the BBC and The Federal Reserve.

3. The government fully regulates the market. In *centrally managed economies,* supply and demand are planned by the government. Businesses are owned by the government.

The extent to which the government stimulates or regulates market forces depends on the political parties in power. *Liberal parties* (the parties right of centre), for example, will look mainly at the effect on competition, employment and the restriction of individual freedom. In their view, market forces and thus competition are important to promote prosperity through low prices combined with quality and innovation. Measures that weaken the competitive position of companies will only be accepted by liberals in the case of very extreme negative externalities.

The political parties to the left of the center acknowledge the significance of competition, employment, and market forces. However, they emphasize the importance of balancing these aspects with social well-being and environmental concerns. These parties advocate for a greater role of government in promoting social interests and implementing policies that prioritize the welfare of individuals and the planet. This commitment is often reflected in their party names, which include terms like "social" and "green" to indicate their specific priorities. Left parties in certain countries, such as the Soviet Union, China, and Cuba, have historically held the belief that market forces had significant negative consequences. In their view, they believed that government intervention was necessary to promote prosperity and mitigate the perceived harmful effects of market forces. These parties subscribed to the ideology known as communism, which is situated on the far left of the political spectrum. According to this ideology, relying solely on regulations was considered insufficient to effectively guide the market towards desired outcomes.

These countries utilize their tax systems to address income disparities and mitigate social inequality, aiming to prevent significant gaps between different segments of society. By implementing progressive taxation and social welfare programs, they strive to create a more equitable distribution of wealth and ensure social cohesion within the country. For example, everyone in The Netherlands has affordable health insurance. America and England are examples of countries that align more towards the right of the political spectrum and follow the Anglo-Saxon model of government. The term "Anglo-Saxon" in this context refers to the predominance of this model in English-speaking countries, including the United States, the United Kingdom, Ireland, Canada, Australia, and New Zealand. America and England are examples of countries that align more towards the right of the political spectrum and follow the *Anglo-Saxon model of government*. The term Anglo-Saxon in this context refers to the predominance of this model in English-speaking countries, including the United States, the United Kingdom, Ireland, Canada, Australia, and New Zealand.[18] This model is 'based on a market economy in which the government has limited its tasks in the field of social security, education and the like as much as possible' (VanDale definition). For example, in the United States, providing basic health insurance for everyone has been a battle between Republicans (right) and Democrats (left) for years.

3.5 Summary

Guiding principles

A guiding principle is a (subconscious) value that underlies people's choices.

Definition of an economic system

A society can adopt various approaches to regulate production and consumption. The foundation of these approaches is the market. However, whether the market operates as a free market, a sustainable market, or a planned market depends on the guiding principles embraced by society. The combination of the market and these guiding principles constitutes the economic system. In most countries, there is a preference for promoting and fostering a market that is as free as possible.

The guiding principles of the free market economy

The free market economy has the following guiding principles:

1. People have the *freedom to serve* their own *interests* When individuals act in their own self-interest, it ultimately leads to a better life for everyone. This concept suggests that the pursuit of individual interests, within the framework of freedom, operates like an invisible hand. It is through this invisible hand that individuals, without intending to do so, contribute to overall prosperity.
2. Productivity is a key factor in combating poverty, and it can be effectively increased through the process of labour specialization. When individuals specialize in specific tasks or trades, they develop expertise and proficiency in their chosen areas. As a result, they are able to produce goods or provide services more efficiently and at higher quality compared to those who have to handle a variety of tasks on their own. This specialization not only enhances productivity but also creates opportunities for trade, as individuals with different specializations can exchange their goods or services.
3. *Free trade*: Export and import must be possible without restrictions. The idea is that if countries are economically dependent on each other, this will lead to less conflict and war. Countless free trade zones have arisen from this idea.
4. *Private ownership*: movable and immovable property belongs as far as possible to private legal persons (individuals or organizations). Private ownership works well in a free market economy because it is consistent

with the principle of self-interest. By doing business you increase your property and thus your wealth. The principle encourages people to work because the result of this work serves their own interests.

5. *Monetary value drives our activities.* In order to thrive in a free market economy, producers must offer goods or services that hold monetary worth and appeal to consumers. This monetary value is, therefore, an important guide in the choices made by individuals and companies. Money directs where we invest our energy. As a result, we tend to allocate less or no money to things that may be valuable but are less or not financially profitable.

6. *Who pays, decides.* In the market economy, companies mainly use wage labour. Wage labour creates a dichotomy between people (employer and employee) and leads to a majority (employees) that performs labour to produce a surplus for a minority (employers).

7. *People need to be motivated.* Economic theory assumes that the employee's attitude is opportunistic and lazy. Opportunistic means that the employee is not necessarily focused on the company's interest but on his or her own interest. This fits with the free market theory: the market works because everyone acts in their own interest. Doing the least possible, while still receiving money is therefore the best strategy for employees. To combat this attitude, employers implement motivational strategies.

8. *Economic growth* is important and wealth at the top naturally trickles down to the bottom of society. All the value added by companies and individuals collectively contributes to the GDP, which serves as a measure of economic growth and prosperity.

The role of government in a market economy

The role of the government varies depending on the political parties in power. There are three main variants: a fully free market economy, a mixed economy and a centrally planned economy. All European countries have a mixed economy.

Notes

1. Sources used to describe the work of Adam Smith include A documentary on Adam Smith, www.youtube.com/watch?v=V6S6pMsKzlI&index=2&list=PLhgPVo-tcOhvBUxttXoZVJ4DJcoHvjakU&t=768s and a brief explanation by Professor Lyne Kiesling: www.youtube.com/watch?time_continue=68&v=DTDUzd6_6Vw. Both accessed October 2, 2022.

2. Legal forms of business. Accessed September 21, 2022, from https://www.kvk.nl/english/starting-a-business-in-the-netherlands/legal-structures-of-business/.
3. Communist countries. Accessed October 2, 2022, from https://www.worldatlas.com/which-countries-are-still-communist.html.
4. Lecture 3: Putting a price tag on life. Accessed October 2, 2022, from http://justiceharvard.org/lecture-3-putting-a-price-tag-on-life/#1477504398584-5d76ef29-aab0.
5. There are several internet resources that can tell you more about (the work of) Karl Marx such as this one: Karl Marx, Paris Manuscripts. Accessed October 2, 2022, from https://humanistischecanon.nl/venster/sociale-bewegingen/karl-marx-parijse-manuscripten/ and https://www.socialistrevolution.org/who-were-karl-marx-and-frederick-engels.
6. Economic update: The contributions of Karl Marx. Accessed October 2, 2022, from http://economicupdate.libsyn.com/economic-update-the-contributions-of-karl-marx.
7. Now there are conceivable situations in which you as an employee still have a considerable bargaining power. The employer usually has more power because there are a hundred others on the doorstep before you (Marx calls this the 'reserve army'). But in unique situations there are only a few, and the roles reverse: the employer becomes more dependent on you than you are on him. For his ten other companies! However, much work is quite simple, especially also due to industrialization, which makes the worker mostly dependent on the employer.
8. Trade union. Accessed September 17, 2022, from https://www.britannica.com/topic/trade-union.
9. Labor movement. Accessed October 2, 2022, from https://www.history.com/topics/19th-century/labor.
10. Economic update: Understanding Marxism (premium episode). Accessed October 2, 2022, from http://economicupdate.libsyn.com/understanding-marxism.
11. Secretly, we're all communists. For centuries. All day long. Accessed October 2, 2022, from https://decorrespondent.nl/4212/stiekem-zijn-we-allemaal-communisten-al-eeuwen-de-hele-dag-door/161930340-c837c928.
12. Why work doesn't make you happy. Accessed October 2, 2022, from https://www.theschooloflife.com/amsterdam/blog/waarom-werk-niet-gelukkig-maakt/.
13. The best stats you've ever seen | Hans Rosling. Accessed October 2, 2022, from www.youtube.com/watch?v=hVimVzgtD6w.
14. Growth. Building jobs and prosperity in developing countries. Accessed October 4, 2020, from www.oecd.org/derec/unitedkingdom/40700982.pdf.
15. Can we do without economic growth? Accessed October 2, 2022, from https://economie.rabobank.com/publicaties/2014/november/kunnen-we-zonder-economische-groei/.

16. De Nederlandsche Bank has produced a lesson on the 2008 crisis. Consulted on 17 September 2022, from https://adoc.pub/van-kredietcrisis-tot-recessie.html.
17. Neoliberalism. Accessed October 2, 2022, from https://www.britannica.com/topic/neoliberalism.
18. What is an Anglo-Saxon economy? Accessed September 17, 2022, from https://www.worldatlas.com/articles/what-is-an-anglo-saxon-economy.html.

Literature

Block, P. (1993). *Stewardship. Choosing service over self-interest*. Berret-Koehler Publishers.

Corbey, M. H. (2010). Agent or steward? On human view and management control. *Monthly Journal of Accountancy and Business Economics, 84*(9), 487–492.

Deci, E. L., Olafsen, A., & Ryan, R. (2017). Self-determination theory in work organizations: The state of a science. *Annual Review of Organizational Psychology and Organizational Behavior, 4*, 19–43.

Gagné, M., & Deci, E. L. (2005). Self-determination theory and work motivation. *Journal of Organizational Behavior, 26*(4), 331.

Georges, A., & Romme, L. (1999). Domination, self-determination and circular organizing. *Organization Studies, 20*(5), 801–832.

Lucassen, J. M. W. G. (2013). *The history of work and labour*. ISHH Paper. https://iisg.amsterdam/files/2018-01/outlines-of-a-history-of-labour_respap51.pdf

Marx, K. (1867). *Capital*. Downloaded as an e-book from Marxist.org

Rigby, C., & Ryan, R. (2018). Self-determination theory in human resource development: New directions and practical considerations. *Advances in Developing Human Resources, 20*(2), 133–147.

Smith, A. (1784). *An inquiry into the nature and causes of the wealth of nations*. MetaLibri.

van den Broeck, A., et al. (2009). The self-determination theory: Qualitative motivation in the workplace. *Behavior and Organization, 22*(4), 316–334.

Wilkinson, R. G. (2005). *The impact of inequality: How to make sick societies healthier*. Routledge.

4

Guiding Principles for a Sustainable Economy

In Chapter 3, you learned about eight guiding principles. Together with the behaviour of customers, shareholders, companies and a government, they form the economic system. A transition to a sustainable market requires both different behaviour of the players and different guiding principles. Each guiding principle from the free market economy can be replaced by a sustainable alternative. This chapter focusses on these sustainable guiding principles and is similar in structure to Chapter 3. In Chapter 5 you will learn more about sustainable behaviour of the market players.

4.1 Acting Responsibly for Greater Well-Being

The detrimental impacts of market forces highlight one important thing, and that is the absence of responsibility. Are you really completely free to satisfy your own interests? Can you enrich yourself at the expense of the earth and your fellow human beings? Should you take responsibility for the other or are you waiting for a government to impose it on you? By using the words 'must' and 'may' we enter the field of ethics. Of how things should be done, of what is right and wrong. This moral finger-pointing often triggers resistance in many people. Interfering with others' behavior is a sensitive issue, as personal freedom is highly valued. This trend is evident in Europe, where many individuals have distanced themselves from the churches they were raised in. The notions of sin and guilt have been replaced by a pursuit of freedom and happiness. When it comes to setting boundaries, governments can regulate them through legislation.

This ethical responsibility for the earth and our fellow human beings was given a new name in the 1980s: *sustainability*. An important step in this was the Brundtland-report, which advocated sustainable economic development, defined as 'development that meets the needs of the present without compromising the ability of future generations to meet their own needs'. A famous phrase, learn it by heart!

The term "sustainable" has often been associated with various other concepts, ranging from biological and organic to alternative, crunchy, and treehugger. For years, people who cared about the earth and their fellow humans were ridiculed by many other people and put in the alternative corner (and thus not to be taken seriously). They were often seen as outcasts, believing in the power of organic tea and tofu to save the world. On the other hand, young urban professionals, such as bankers, were regarded as the epitome of success.

But something has changed in recent years, especially after the economic crisis of 2008–2009. With the recession and increasingly alarming reports of climate change, sustainability has moved away from its boring image, from something that is nice but not necessary. People are making themselves heard more. For example, in 2008 a group arose that questions economic growth: the DeGrowth movement. They challenged the current economic and social paradigm, characterized by 'faster, higher, further', and strive for an economy that puts people's well-being first. In 2011, the Occupy movement protested against an unstable economic system with harmful effects on people and the environment.

In 2019, there was a global surge of young people raising their voices, spurred by Greta Thunberg from Sweden. Following her summer break in 2018, Greta decided to take a stand for the climate by going on strike. This movement, known as climate truancy, received permission from schools in The Netherlands for a day of protest. Greta's powerful message questioned the purpose of attending school when the future is endangered by climate change. She quickly rose to fame, and in 2019 she was allowed to address world leaders at the United Nations Climate Conference (the COP24, where COP stands for *Conference of the Parties*).

Sustainability is an alternative to morality or norms and values, but with the same tenor: there are limits to freedom, self-interest and growth. The earth and our fellow humans form that limit. Or, as described in 1789 in Article 4 of the Declaration of the Rights of Man and the Citizen: 'Liberty consists in being able to do anything that does not harm another'. Even Adam Smith wrote a book about morality, next to his study on markets! Therefore, before delving into the intricacies of sustainability, it is essential to grasp the concept of moral behavior.

Responsible Self-Interest Through Ethical Awareness

The more educated reader may have raised an eyebrow at Smith's assertion that serving of one's own interests is central to the free market. *The Wealth of Nations* is not Smith's only book. It was preceded by *A Theory of Moral Sentiments* (1759).[1] In this book, Smith investigates whether people are more focussed on their own interests or whether they are also benevolent towards others. Whereas other philosophers tried to answer these questions mainly on the basis of their own thoughts, Smith—fitting in with the critical spirit of the time—carefully observed human behaviour. He found that people learn what is right and wrong in interaction with each other, and that they adjust their behaviour accordingly. When a child takes away their sibling's toys, causing them to react with distress, their mother intervenes to teach an important lesson: taking things from others without permission is not acceptable. Through this guidance, children begin to understand the importance of self-control and respecting the possessions of others. Over time, they learn to internalize this norm and develop a sense of conscience that guides their behavior. The mother's initial intervention serves as a catalyst for the child's moral development, enabling them to navigate social interactions with greater empathy and consideration. Smith calls that conscience the impartial spectator, as if someone is watching inside your head when you have to choose between good and bad.

A moral compass for one's behaviour is therefore developed through education and contact with other people. Smith calls these informal norms and values, in addition to the formal ones, which are laid down in laws by the government or by religions.

In addition to these formal and informal values, Smith also sees a third basis of moral behaviour: empathy. When people see the suffering of others, they feel that suffering. Empathetic people become happy when they can make other people happy. This empathy ensures that they do not want to harm others, which is the basis for the well-known expression 'Do not do unto others what you would not want done to you'.

> **Example**
> **Roots of Empathy**
> It is increasingly recognized that the development of empathy is important for a sustainable functioning society. The organization Roots of Empathy is committed to this development. In the short term, Roots of Empathy focusses on increasing the level of empathy for other human beings, resulting in more respectful and caring relationships and less bullying and aggression. The ultimate goal is to create a world in which everyone feels responsible for the well-being of others and the planet.
>
> *Source* Roots of Empathy

Informal norms and values, laws and religion and an innate tendency not to want to hurt others will thus constrain the maximum pursuit of self-interest, according to Adam Smith (Fig. 4.1).

From this theory, it is easy to understand why Smith became a proponent of the market in which self-interest is given free rein. After all, as long as the market is embedded in the three elements Smith mentioned, it provides an excellent means of increasing everyone's prosperity.

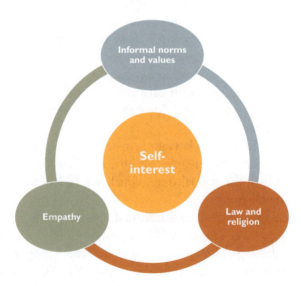

Fig. 4.1 Smith argues that self-interest maximization is constrained by laws and religion, informal norms and values and empathy

Smith is not alone in considering the balance between one's own interests and the interests of others. This issue is part of the ethical philosophy that seeks to answer questions such as: What is good behaviour? When do I behave responsibly? When am I a good person? Several ethical theories have attempted to answer these questions: utilitarianism, duty ethics and virtue ethics.

The ethics of utility (utilitarianism)

In utilitarianism, also known as consequentialism, the morality of an action is determined by its consequences and the overall impact on the common good or welfare. Philosophers who promoted this movement are Jeremy Bentham (1748–1832) and John Stuart Mill (1806–1873). Utilitarianism focuses on the outcomes or consequences of actions, and it aims to maximize overall happiness or utility in society. This perspective often involves making decisions based on a cost-benefit analysis, where the benefits and harms resulting from different choices are carefully considered. A famous example is that of the shipwreck in which four people survive. After days of floating, they are near death. The captain decides that the youngest, and weakest of the four, may be killed so as to increase the chance of survival for the other three.[2] From a utilitarian perspective, this was the right decision: three people stay alive and only one dies. Even in modern times, people often reason on the basis of consequences. For example, during the corona crisis, some people questioned whether the social constraints outweighed the benefit of protecting our elderly. The story of Ford Pinto in Chapter 3 was another example of this ethical movement, as the monetary value of lives was weighed against the costs of repair.

The utilitarian idea of *the greatest good for the greatest number* can help find the right balance, for example, in the case of comparable values or between people in the same group. But what if the production of clothes in Bangladesh that leads to the happiness of all the Western girls who can buy a dress cheaply at Primark has to be weighed against the poor working conditions of the Bangladeshi textile workers and against the chance that they will die because of working in poor buildings? If there are more happy girls than suffering people in Bangladesh, is producing textiles in Bangladesh right? From this ethical point of view, yes, but most people feel that there is something wrong here. Thinking from the perspective of utility for most people does not always lead to the right action.

The ethics of duty

Immanuel Kant (1724–1804) rejected the idea that morality is only about weighing the consequences. He argued that each of us has certain rights and duties that always apply, even if the resulting actions do not yield the most benefit. Kant formulated two criteria to judge whether an action should be regarded as a duty:

1. A human being should never be approached merely as a means to an end. In concrete terms, this means, for example, that values such as honesty, justice and the human dignity must be respected.
2. An action can be seen as a duty if it can be a universal law for all of us. Familiar examples are 'you must not kill' and 'you must not lie', because if everyone does, society as a whole will be worse off. These are the informal norms and values that Smith mentioned.

Go back for a moment to the shipwreck, where one person was killed to save the other three. Had the captain reasoned from Kant's philosophy, he shouldn't have done this. After all, he himself would not have wanted to be killed either. Thinking about the poor working conditions in Bangladesh, this is morally wrong, even if those poor conditions lead to cheap clothing that makes other people happy. We have a universal duty to offer these children a good childhood. Just as there was a group of people during the corona pandemic who felt we should protect the lives of the elderly at all costs. You cannot balance the right to live against another interest, such as what it costs to keep these people alive.

The ethics of virtue

Virtue ethics differs from other ethical frameworks as it focuses not on duties, but on the aspiration to be a virtuous and admirable person. Do you know someone who is wise, sensible and modest? Or do you know someone else who you admire for the values he or she pursues? And do you have the ambition to be a good person yourself? Aristotle (384–322 B.C.) laid the foundations for virtue ethics. He stated that people are inclined to goodness and become most happy by developing certain virtues. He describes four virtues, summarized in a publication by Senternovum, a former agency of the Dutch Ministry of Economic Affairs (edited)[3]:

> Wisdom, Sense of reality, Courage and Commitment. According to the classical tradition, only those who have made these values their own, and thus made them virtuous, are 'whole and healthy' human beings. Wisdom enables

people to assess a situation well, to make good judgements and choices, and to act accordingly. But insight into what is right to do does not in itself guarantee that what is right is actually done. The causes of this are many, but can be summarized roughly as fear and temptation. To overcome these, two other virtues are important. Courage (the awareness of one's own inner strength) is needed to overcome fear, and a Sense of reality (or moderation if you wish), which helps us to resist temptation. And finally, Commitment, a well-developed sense of justice, is an indispensable quality for a virtuous person.

Even Adam Smith ends his first book with virtue ethics, describing the character of a truly virtuous person, in line with Aristotle's classical values.

Virtue finds its translation in the concept of the purpose economy. According to Aaron Hurst, author of the book *The Purpose Economy* (2017), it is the next phase in our society. The purpose economy is about finding your own place on earth. You find it by asking the following questions: what do I enjoy doing (1), what competencies do you have (2) and what does the world need (3)? Once you have answered these questions, you can determine from which position you can monetize that (4). In this way you not only take care of your own well-being and happiness but also of the world around you.

> *Be a contribution* write Zander and Zander (2002) in their book *The Art of Possibility*. They describe how a musician translated this into her work: '*I am now able to use the possibility that my every act can affect the world to communicate with people in such a way so that a wave of inspiration and happiness can flow throughout the world. I know now that music is not about fingers or bows or strings, but rather a connective vibration flowing through all human beings, like a heartbeat. It is my job and ambition to keep that invisible and easily-cut lifeline free and supported in all parts of life*' (Zander & Zander, 2002: 62).
>
> Do you know what is or will be the meaning of your life? What makes your study or work so important that you want to give a large part of your life to it? What values do you find important? And are they only values that improve your life for yourself, or are they values with which you also contribute to a better world? Henry Ford, the founder of the Ford car company, had these values:
>
>> Ford, a quintessential capitalist that he was, thought otherwise: 'I don't believe we should make such an awful profit on our cars. A reasonable profit is right, but not too much. I hold that it is better to sell a large number of cars at a reasonable small profit (…) I hold this because it enables a larger number of people to buy and enjoy the use of a

> car and because it gives a larger number of men employment at good wages. Those are the aims I have in life'. (Nevins & Hill, as cited in Adams et al., 2009)

Tip
If you want to find out which values are important to you, take a look at the site of World Values Day. They have made a list that can help you become aware of what you find important.

Sustainability: utility, duty and virtue

If Adam Smith's ideas had been fully realized, it is argued that we might have avoided some of the negative consequences associated with the market. Nearly two and a half centuries later, the complexity of our reality has unfortunately become more apparent. This raises questions about our past knowledge and foresight. Did we anticipate the significant environmental impact of CO2 during the early days of the industrial revolution? Were we aware of the future challenges posed by plastic waste when we first introduced plastic products? Did clothing retailers understand the extent of poor working conditions in manufacturing plants? And how long have we recognized the ethical responsibility we have towards animals and their well-being? These questions prompt us to reflect on whether our past actions stemmed from a lack of wisdom or if our present understanding represents a progressive moral insight.

Sustainability is a modern attempt to make a virtuous, right life concrete. It is a *moral retinue around self-interest and the compelling hand of the market* and can be reasoned from all three ethical theories:

- Sustainability makes sense in a utilitarian framework if the benefits outweigh the costs. For example, solar panels lead to lower energy costs, and infrastructure costs fall if CO_2 emissions are reduced. After all, no dikes would have to be built if sea levels didn't rise.
- When examining sustainability through the lens of duty ethics, greater emphasis is placed on universal duties and values. For instance, while it

might be financially advantageous to employ children in clothing production due to lower labor costs, this practice violates children's rights and is consequently considered unacceptable through this ethical lens.
- The compelling hand of the market keeps the card designer from printing cards on recycled paper (because it is less beautiful and more expensive). However, if they want to do good for the earth and be a virtuous person, they will take that step anyway and settle for a lower profit margin.

Sustainable encompasses the concept of long-term viability and the ability for something to persist over time. When considering sustainability, one can easily relate it to nature. For example, practicing sustainable forestry involves ensuring that trees are replanted or regenerated, with the goal of minimizing the number of trees cut down since they play a crucial role in absorbing CO_2. Similarly, sustainable fishing involves managing fish populations in a way that allows them to thrive and reproduce, thereby maintaining their strength and abundance. In a broader sense, sustainability means that we should not exploit Earth's resources beyond their capacity to regenerate and provide for both the present and future generations.

But this definition does not always apply. Using children for labor cannot be justified or considered sustainable, even if their basic needs for food and rest are met. Using children for labor cannot be justified or deemed sustainable, even if their basic needs for food and rest are met. The question of whether they can withstand such labor in the long term raises ethical concerns. From the perspective of duty ethics, this practice is considered immoral. The immorality of child labor is firmly established in the International Convention on the Rights of the Child, which was drafted and unanimously adopted by the United Nations in 1989. Furthermore, the Universal Declaration of Human Rights, formulated by the UN in 1948, emphasizes that certain rights of individuals, including children, take precedence over any other values or utilitarian considerations. These treaties recognize the inherent rights of individuals and prioritize their well-being over any short-term gains or economic benefits.

Sustainability has two main elements: the earth and people. John Elkington related these concepts to business in his book *Cannibals with Forks* (1999). In this book he defines the *triple bottom line*, consisting of people, planet and profit (the three Ps). In this framework, financial profit should be considered on an equal level with social responsibility and ecological protection (Fig. 4.2).

In 2017, Kate Raworth presented another powerful model: the donut economy. On her website she says the following[4]

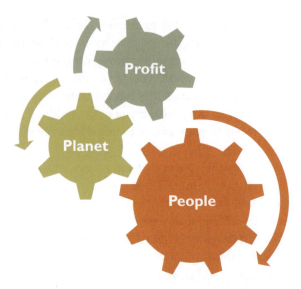

Fig. 4.2 People, Planet, Profit: sustainable companies produce goods and services that generate profit for themselves, for the earth and for their fellow human beings

> Humanity's 21st century challenge is to meet the needs of all within the means of the planet. In other words, to ensure that no one falls short on life's essentials (from food and housing to healthcare and political voice), while ensuring that collectively we do not overshoot our pressure on Earth's life-supporting systems, on which we fundamentally depend—such as a stable climate, fertile soils, and a protective ozone layer. The Doughnut of social and planetary boundaries is a playfully serious approach to framing that challenge, and it acts as a compass for human progress this century.

Raworth represented this challenge with two circles, making the model resemble a doughnut. The inner circle, the social basis, indicates our basic needs for a dignified existence and includes food, housing, health and resilience. If these are met, then the human being is in the safe zone, the place where it is good to live and where there is room for the fulfilment of desires instead of needs. The outer circle then indicates where those desires reach a limit. If we go through the ecological ceiling, we ask more of the earth than she can give us (Fig. 4.3).

John Elkington and Kate Raworth are not the sole voices discussing sustainability. The concept of sustainability has a rich historical background. Prior to the 1960s, the primary concerns revolved around the increasing population and the potential strain it would place on natural resources. The English demographer and economist *Thomas Malthus* wrote as early as 1798

4 Guiding Principles for a Sustainable Economy   109

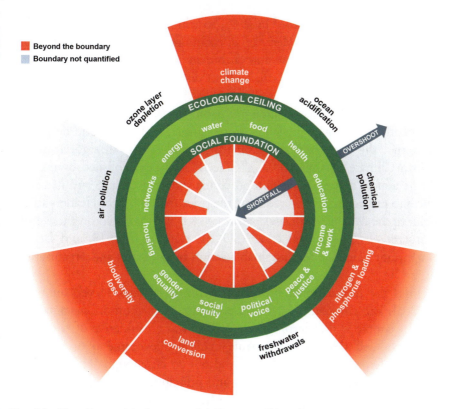

Fig. 4.3 Kate Raworth's donut model (*Source* Wikimedia Commons—DoughnutEconomics)

in *The Principles of Population* that eventually there would not be enough food to feed everyone.

Society's current perspective on the limits of the economy came about primarily through the publication of the *Club of Rome*'s report *Limits to Growth*. This club was founded in 1968 and consisted of a group of scientists and other influential people who were concerned about the depletion of our earth. This report calculates a number of scenarios based on factors that influence human development, for example, population size and resource depletion.

In 1987, the *World Commission on Environment and Development* published its report *Our Common Future*, which was already featured in this book. Their definition (you will find it at the beginning of this chapter) has become the basis for the concepts of sustainability and sustainable development. The United Nations commissioned Norwegian Prime Minister Brundtland to write this report. The report fits the ambitions the UN had already set when it was founded in 1945: 'to achieve international cooperation in solving international problems of an economic, social, cultural or humanitarian nature and to promote and encourage respect for human rights and for fundamental freedoms for all, without distinction of race, sex, language or religion'.[5]

Since then, the UN has created numerous initiatives to put sustainability permanently on the global agenda. For example, in 1992 in Rio de Janeiro they started the *conferences on climate* organized by the Intergovernmental Panel on Climate Change (IPCC).[6] The IPCC was established by the UN with the aim of mapping research on climate, climate change and its consequences to provide advice on reducing the risks. In 2000, world leaders agreed to achieve the *Millennium Development Goals* by 2015. Based on these goals, the *Sustainable Development Goals* (SDGs)[7] were set in 2019, and serve as an important guideline for sustainable action by countries and organizations (Image 4.1).

Besides the international initiatives at the government level, numerous individuals and companies have also taken action to promote sustainability. A good example is the *Earth Charter*.[8] This was an initiative of two world leaders who felt that the agreements made by governments at the 1992 climate conference in Rio de Janeiro did not go far enough. They therefore thought that a document with values and principles for a sustainable world could also be drawn up by individuals. After many drafts and with the input

Image 4.1 The United Nations' seventeen Sustainable Development Goals (*Source* United Nations)

of people from all parts of the world, the Earth Charter Commission came up with a final version in March 2000. It consists of four main principles:

1. respect and care for all forms of life
2. ecological integrity
3. social and economic justice
4. democracy, non-violence and peace.

In the Netherlands, Urgenda is a prominent grassroots initiative, formed by combining the words "Urgente Agenda" (Urgent Agenda). The foundation was established in 2007 by Jan Rotmans and Marjan Minnesma from Erasmus University Rotterdam, with Jan Rotmans serving as a professor at the university. One of the foundation's achievements was taking the Dutch government to court in 2015, demanding that it protect the citizens of The Netherlands from the consequences of climate change. Urgenda felt that the government was not taking sufficient action in addressing sustainability issues. The uniqueness of the case lay in the fact that it was the first time citizens had taken the government to court on such grounds. After all, if you want to achieve something nationwide, you go into politics to get it done. Urgenda's position was vindicated by the court in 2015, but the government chose to appeal the decision and continued to pursue the case in higher courts, despite numerous requests from citizens to refrain from doing so. The Supreme Court ruling, however, upheld Urgenda's claim.[9] The *Occupy movement* and the *youth protests* ignited by Greta Thunberg (discussed briefly at the beginning of this chapter) are other examples of civic initiatives.

Sustainability for companies has led to the emergence of *Corporate Social Responsibility* (CSR). To support companies in making this transition, several knowledge and information centers have been established in various countries that are affiliated with CSR Europe. Their 'European Pact for Sustainable Industry aims at scaling up the impact of individual efforts made by companies, industry federations, and EU leaders towards a Sustainable Europe 2030'.

Through these initiatives, Non-Governmental Organizations (NGOs), individuals, and organizations demonstrate their awareness of the problems we face and their commitment to taking responsibility for them. They reject the guiding principles of an unsustainable society and market, where self-interest is allowed to come at the expense of the interests of the earth and other people. In other words, in a sustainable market economy, the freedom to pursue self-interest is complemented by the principle of ethical responsibility towards the interests of the planet and all living beings. It acknowledges that economic activities should not only prioritize individual gain but also consider the long-term well-being of the environment and society as a whole.

Interdisciplinary Work for Sustainable Solutions

Taking responsibility is indeed closely tied to one's ability to influence the outcome. In a society characterized by labor specialization, individuals often have a more focused impact within their specific area of expertise or domain of work. This means that their ability to take responsibility and make a direct influence may be constrained to their particular field. Let's take a closer look at the example of a communications advisor who wants to promote sustainability within an organization. In order to bring about change, the advisor must first convince others within the organization of the importance of sustainability and the need to align their actions with this vision. Once the organization embraces the sustainability vision, it must be translated into action by various specialists and departments. This might involve collaboration with experts in different areas such as operations, finance, procurement, and human resources, among others. Each specialist plays a vital role in implementing sustainable practices within their respective domains. It is important to acknowledge that such organizational transformations take time. The process of changing mindsets, shifting practices, and aligning efforts across different functions is gradual and requires persistence. Some individuals may resist change and prefer to stick to their familiar ways of working. The result is that many people (try to) do their own work as well as possible, but that the sum of all those partial activities does not lead to sustainable outcomes.

To achieve sustainable results, cooperation between the various professional specializations is necessary. This is called *interdisciplinary work*. Scientists call the problems in which many people from different disciplines and organizations have to work together to find a solution to *wicked problems*. According to Professor Rob van Tulder, founder of the Wicked Problems Plaza, such a problem has four characteristics[10]:

1. The various parties have incomplete or contradictory information about the problem.
2. A diversity of opinions and possible solutions exists between the parties, which hampers reaching a final strategy.
3. The economic costs are too great for the individual parties involved.
4. The different parties have unequal or incompatible desires and interests.

The transition to a sustainable economy is one such wicked problem, as are many related sub-problems.

If we want to tackle wicked problems, the entire system has to go along with it. This change is therefore also called a system change. It is about doing things right, and doing the right things together. That requires others. The guiding principle of labour specialization for higher productivity changes in a sustainable market to the principle of interdisciplinary cooperation for the solution of wicked problems.

Local Trade and International Sustainable Chains

Free trade was the third feature of the free market economy. In Chapters 1 and 2 you learned about the ways in which companies trade internationally and the harmful effects this has. How will this change if we want to create a sustainable market economy?

Protectionism is a contentious yet employed strategy adopted by governments, whereby countries opt to impose import tariffs or quotas. The implementation of protectionist measures by Donald Trump, the President of the United States from 2017 to 2021, caused a significant stir worldwide. Trump's intention behind this approach was to generate more job opportunities within the United States and revitalize the nation's economy. His actions represented a departure from the prevailing post-World War II trend and ran contrary to the longstanding notion of an open global economy that inherently seeks self-regulation and equilibrium.

Besides government responsibility, companies can also assess their offshoring practices. One aspect of *sustainable free trade* is when companies take responsibility for maintaining a sustainable supply chain. In many cases, products and semi-finished goods are produced in low-wage countries and then transported to another country for additional processing. For instance, the journey of cotton in your jeans involves multiple flights and shipments across the globe before it reaches your closet (Image 4.2).

To combat scandals and abuses, NGOs have been set up to promote ethical trade. A good example is the Ethical Trade Initiative: 'ETI was created in 1998 by a small group of visionaries who believed in the power of collective action to make a difference to the lives of workers in companies' supply chains. All corporate members of ETI agree to adopt the ETI Base Code of labour practice, which is based on the standards of the International Labour Organisation (ILO). We work out the most effective steps companies can take to implement the Base Code in their supply chains'.[11] It is one example of companies trying to reduce the coercive hand of the international free market.

Image 4.2 Our products are transported all over the world before they arrive in your home (*Source* Pixabay, plindena)

Secondly, businesses and individuals can promote the local economy by fostering the development of *local production chains*, often referred to as "short chains." This involves establishing closer connections between producers, suppliers, and consumers within a specific region. In the field of food, for example, the Rural Network sees 'a multitude of short chain initiatives, varying from farmyard sales, sales via new farmers' cooperatives, new retail concepts, crates and boxes, farmers' markets and food halls, city farms and consumer collectives'. These are small-scale short chains, reducing complexity and enabling successes to be achieved quickly. Energy is another sector where local chains are becoming fashionable, especially with the rise of solar and wind energy. Instead of importing gas from Russia, residents can now generate their own energy, or do so collectively. Short local supply chains offer several benefits, including a reduction in transportation movements and an increased level of engagement and awareness in the production process. Consequently, consumers are more inclined to make sustainable decisions.One crucial aspect of this awareness is recognizing the working conditions in which products are manufactured, such as the exploitation of child laborers who endure long hours for meager wages. This realization raises an important question: how can one derive satisfaction from purchasing a cheap T-shirt while being fully aware of the hardships faced by those involved in its production? It becomes apparent that such purchases do not contribute to the improvement of these workers' lives, as only a minute fraction of the purchase price actually reaches them.

In a sustainable market, the principle of international free trade has changed into the principle of sustainable international chains and local production and consumption.

Ownership and the Sharing Economy

In Chapter 3, you learned that ownership is an essential part of market forces. But how logical and necessary is ownership? There are two perspectives. First, ownership can be positive for a sustainable economy. Remember the tragedy of the commons? If more people are allowed to use a common asset, there is a good chance of overexploitation. Misuse, overexploitation and destruction of movable or immovable property is thus more likely if you don't own the property, which fits within the economic theory in which everyone looks after their own interests.

Organizations have arisen to protect nature, such as Greenpeace, National Trust and the World Wide Fund for nature. Some of these organizations buy land for conservation, thereby protecting it through ownership. The government makes an important contribution to financing these organizations. The government is also responsible for designating national parks. Another example where people try to exert influence through ownership is by buying shares in companies. These shares give them access to the shareholders' meeting, and thus the opportunity to ask questions about sustainability.

A second perspective in the context of a sustainable economy is the shift from individual ownership to a focus on shared and collaborative consumption. This transition involves moving away from the traditional model of owning goods towards a more sustainable approach centered around the concept of use. Instead of everyone owning a car, in a few years we will be driving around in self-driving cars that you can call up with an app. Instead of bulbs, we will buy light hours. The rise of shared scooters is another good example. The big advantage is that this gives manufacturers an incentive to make the most sustainable product possible. The longer a product lasts, the more they can earn from it. As soon as the product has to be replaced, the manufacturer takes it back for reuse (in whole or part). The transition from possession to use fits into a world view in which it is not possession, but rather experiences that create happiness.

In the case of a use economy, there is still a central manufacturer or entity that owns the product, but the emphasis shifts from individual ownership to the utilization and access of goods and services. But what if we share ownership? Instead of each individual owning their own lawnmower, an alternative in the sustainable economy is to collectively purchase a high-quality lawnmower as a neighborhood and share its use. This concept has been coined as *the sharing economy*, which promotes the idea of collaborative consumption and resource sharing. The advent of the Internet has significantly facilitated the growth of the sharing economy. Online platforms provide a digital marketplace where owners can offer their possessions or resources for sharing with others. These platforms are commonly known as *peer-to-peer platforms* because they facilitate direct sharing and transactions among individuals without the need for intermediaries or traditional companies. It is lucrative for three parties. Firstly, the platform operator or manager can generate revenue through various means, such as charging a commission or transaction fee for facilitating the sharing transactions or by displaying advertisements on the platform. This revenue helps sustain the platform and supports its ongoing operations. Secondly, the lender or owner of the shared item can earn income by charging a small fee or rental amount to the borrower for the use of their item. This allows the lender to monetize their underutilized assets and generate income from resources that would otherwise remain idle. Lastly, the borrower or user of the shared item benefits by avoiding the need to purchase the item outright. By renting or accessing the item through the platform, they can enjoy the temporary use of the item without the financial burden and responsibility of ownership. Such platforms have been developed in many sectors. There are also *peer-to-peer marketplaces*. This is not about sharing, but about swapping or selling. A well-known example is Etsy (selling handmade items) and Vinted.

Such marketplaces are often classified as part of the sharing economy, but this is not justified. After all, existing property is not being shared but sold. The platform is no more than a new outlet for buying and selling stuff. Many companies market their products this way. In that case, these platforms are called *business-to-peer platforms*: companies sell their products to consumers as if they were one of the *peers*. However, that can occur on the same platform as the peer-to-peer marketplace. Many large companies change the true nature of the sharing platform in this way.[12] A critical note, then, is that these platforms give room for people to turn the good deed into an economic business model. Instead of getting food from your neighbour, someone else offers a fresh meal every day for a certain amount of money. Not because there are some leftovers, but because it is a business model for them. There is no longer any question of sharing. It leads to an economization of virtues such as kindness and hospitality, says philosopher Byung-Chul Han.[13]

A sustainable perspective also has implications for *the ownership of organizations*. In Chapter 3 you learned that ownership and the decision-making power go hand in hand. If the power lies with shareholders who want as much return on their investment as possible, decisions will focus on maximizing profit. This does not lead to the most sustainable decisions. Price, innovation and quality will often be more important than sustainability. In a sustainable economy, multiple stakeholders, including employees, become co-owners of the company. This is more likely to lead to a balance between profit, people and the earth. This can be a *cooperative*, but that is not the only form available. Employees and other stakeholders can also receive shares in a private or public limited company. If the owner arranges this co-ownership for the employees, it's called *employee participation*. This can be beneficial, as 'employee share ownership and profit sharing can increase worker pay and wealth and broaden the overall distribution of income and wealth, a key ingredient for a successful democracy' (Blasi et al., 2018).

A third form that has been employed for many years is steward ownership or associations. In organizations that adopt this structure, neither the shareholders nor the founder assume ownership of the organization. Instead, the founder invests the shares in a foundation, effectively making the foundation the owner, and responsible individuals are appointed to oversee the management of the organization. In this setup, no single individual holds ownership, and both the founder and employees are employed by the company. Consequently, there are no shareholders solely focused on profit maximization, nor owners seeking personal enrichment from the company. Notable examples of companies that have implemented this approach include Bosch and Stichting Sleipnir.

> A Dutch example is the Sleipnir Foundation, founded in 1983 by Koos Bakker, founder and owner of Odin, a wholesaler of organic and sustainable products. Touched by the anthroposophical ideas of Rudolf Steiner, Koos Bakker decided that his organization should not be guided by his interests as owner, or—in the case of an IPO—by the interests of shareholders. He neutralized the personal interest by investing the ownership in a foundation. There are now ten affiliated organizations, all of which have the limited partnership as their legal form. 'With the Sleipnir Foundation as joint limited partner, entrepreneurs from ten companies work together without accumulating private assets. The company assets are contributed by the Sleipnir Foundation and serve as a vehicle for the entrepreneurs involved. They do not consider their companies as personal property and therefore can focus

> on delivering high quality products or services that meet the needs of their customers. In doing so, they are working on an economy that is socially sound'.[14]

Shared ownership also regulates control in organizations. You will learn in Sect. 4.2 what this means for the subordinate position of employees. In Chapter 5 you will learn how shared ownership can be organized in concrete terms. In summary, the principle of ownership from Chapter 3 transforms into the principle of protected ownership, use and/or shared ownership.

Multiple Values Guide Our Activity

In Sect. 3.3, the emphasis was placed on the dominance of money within the market. However, in a sustainable economy, a different approach is pursued, one that encompasses *multiple values*. Our actions are no longer solely driven by financial gain but are also guided by a sense of purpose. We acknowledge that there may be instances where we prioritize other values over monetary gains, such as contributing to a healthier environment or assisting others in need.

An example of a product that embodies multiple values is free-range chicken, as opposed to factory-farmed chicken. Free-range chicken not only provides a source of food but also promotes animal welfare and environmental sustainability. Similarly, choosing clothing made of sustainable fabrics by adult workers instead of opting for cheap shirts produced by exhausted children in poorly maintained factories in Bangladesh is another example of prioritizing multiple values. In these examples, the distribution of profits is shared equitably among all the production partners, reflecting a commitment to fairness and ensuring that all contributors are appropriately compensated for their efforts.

Such a product therefore has a *fair or true price*. A true price incorporates all costs, including those normally excluded from the price, such as combating CO_2 emissions and waste processing. In the current economic system we pay these costs as a society (also called externalities). A fair price also means that the costs have not been kept down at the expense of others or the planet. To come back to the production of clothing: if the labour costs of the textile workers are included in the price, but they earn so little that they have to work 12 hours a day, 7 days a week, it cannot be considered a fair price. This is especially true when the clothing is sold at a high price, resulting in substantial profits for the producer. The difference between an

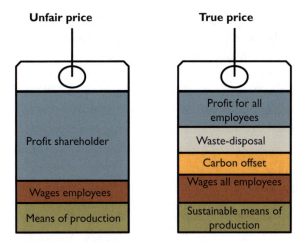

Fig. 4.4 The difference between an unfair price and a fair price (*Source* Picture created by myself, courtesy of the four students)

unfair and a fair price is simplified in Fig. 4.4, which shows negative externalities are included in the price, and wages and profits are divided fairly among all production partners.

This assumes that the price for the consumer does not change, instead extra costs come at the expense of profit. Of course, this is not always the case; sometimes the profit is already so marginal that the company is forced to raise the price. And then the sustainable product struggles because of the coercive hand of the market, where consumers often still choose the lowest price.

In addition to consumers opting for a fair price, a sustainable economy also requires investors (e.g., banks, pension funds and private investors) to provide capital to sustainable businesses. Investing in sustainable initiatives is called *impact investing*. These investments often yield lower returns than, for example, investments in the arms industry. Did you know that in 2007 it was an eye-opener that the Dutch pension fund PGGM invested the money it managed in cluster bombs, land mines and child labour? It provided pensioners with a good pension, but at what cost? The ethical awareness of many investors has changed somewhat since then, but sustainable investments are still uncommon. For example, by 2021 PGGM will have invested only 18% of the total assets under management in companies and projects which contribute to sustainable development in accordance with the UN's Sustainable Development Goals.[15]

Oikocredit demonstrates that sustainable investment can pay off for investors and recipients alike. This organization was founded by an alliance of churches, writes Arjan Broers in his book *Money and Good*. In the late 1960s, the churches wondered how they could invest their money as ethically as possible. In 1975, the forerunner of Oikocredit was born: eighty churches put one million dollars into their own Ecumenical Development Cooperative Society. By 2019, Oikocredit is investing over 1 billion euros a year in three sectors: microfinance for entrepreneurs unable to borrow from a bank, agricultural cooperatives and small-scale energy projects with a major impact.

> Through local partners, Oikocredit extends microcredits to entrepreneurs in Africa, Asia and Latin America. The local partner has direct contact with the entrepreneur. They know the circumstances and are close to the microfinance recipients. An employee of the partner often jumps on a moped to drive to an inaccessible area. There they have personal contact with the microcredit entrepreneur.[16]

> **Tip**
> Grameen Bank is another example of the impact of sustainable financing. Founder Muhammad Yunus received the Nobel Peace Prize in 2006 and talks in a video about his remarkable way of working.[17]

In a sustainable market economy, the market is not guided by financial values alone, but by multiple values.

4.2 People Are Equal Partners in Production

In Chapter 3, you learned about the position of humans in the current economic system: humans are used as means of production. You read about the theory of one of the most important critics of the current capitalist system, Karl Marx. Exploitation, dissatisfaction, alienation and poverty are major negative externalities of this system. Piketty (you read about him in Chapter 2) shows that financial inequality grows, and with it social inequality in society. Marx argues that this system too, like slave labour and feudalism, is ultimately unsustainable.

How can we organize our activities in organizations, without these negative externalities? Is there a solution that ensures that people are not a means of production, but are seen as human beings? Can organizations be created where individuals no longer experience exploitation, dissatisfaction, and alienation? What defines fair working conditions? Is it possible for people to be seen as ends in themselves rather than mere means?

In recent decades, much attention has been paid to people's position in hierarchical organizations. Numerous books have been written about this, especially in organizational psychology and business administration. As a solution for the mind-numbing work on the assembly line, theories and practical manuals for task broadening (let people perform multiple tasks instead of one), task enrichment (make tasks more interesting) and task rotation (let people do something else for a change) were developed. A press button to stop the assembly line was placed so that people could once again be in charge of the machine instead of the machine dictating their pace. Investments were made in motivation, personal development and team building. In modern times, employees have the option to enroll in mindfulness courses to better manage work-related stress. Managers can undergo leadership training to enhance their ability to guide and supervise their employees. In an attempt to address ownership-related concerns, numerous companies have implemented bonus systems. It is often the owner who determines the bonus and sets the required turnover.

Few companies, however, have done anything about the employee's subordinate position in the organization. Remember Chapter 3 talking about improved working conditions for slaves? The master gives them better food, working and sleeping hours are established, and even the place where they sleep is refurbished. Corporal punishment is banned. Perhaps there were slaves who were very satisfied with this situation and their productivity increased. But the fundamental principle that humans were property was not questioned. Marx extends this reasoning to the position of people in the hierarchical organization (he calls this organization by the person: the capitalist). Even if you improve the working conditions of the employees, and quite a lot of people are satisfied with the job they have, Marx argues that it is fundamentally wrong to let people work for you and not let them share in the profits. In the capitalist system, people are underpaid so that in the end there is a surplus value/profit left for the owners of the company. Capital still dominates decision-making in many organizations: 'he who pays the piper calls the tune'. The strategies you learned in Chapter 1 (managing employee productivity and keeping salary costs low) are only possible in hierarchical relationships.

> **Tip**
>
> Is hierarchy and subordination in organizations fine because many employees are happy, or is something fundamentally wrong? To arrive at your own vision, you can fall back on what you have learned about ethics. If you argue that the system is not so bad because most people are happy with it, you are a utilitarian. Marx was not a utilitarian, he thought more along the lines of duties and universal truths. According to Richard Wolff, Marx's truths were the values from the French Revolution (1789–1799): liberty, equality and fraternity. The French Revolution came about because the majority of the population was subservient to the king and the nobility. This majority struggled because of low income and high taxes. The feudal tax system was abolished by the Revolution.

Equality Between Production Partners: Who Contributes, Decides

In opposition to hierarchy and subordination stands *equality*. Equality means acknowledging that every human being has equal value. One human life is not more important than another, even if you are richer, smarter or have an important position in society. Equality is different from sameness; after all, people differ in many ways. But that does not diminish the value of every human being. Equality was one of the three values recognized as a human right in the French Revolution. This value was adopted in 1948 (together with liberty and fraternity) in article 1 of the Universal Declaration of Human Rights:

> All men are born free and equal in dignity and rights. They are endowed with reason and conscience, and should behave towards one another in a spirit of brotherhood.

The pursuit of equality has a long history. Slaves fought for their equality and freedom, the lowest classes rose up against the power of the king and nobility, and women had to wait a long time before they were acknowledged as equal to men (Image 4.3).

Equality is at the heart of a sustainable economy, because it puts the value of a human life first. It means that a human being cannot be seen as a resource and cannot be exchanged for other values, such as lower costs or more production. Equality is at the heart of the UN SDGs and is named in Goal 10, *Reduced Inequalities*.

Image 4.3 Reducing inequality is one the SDG's (*Source* United Nations)

Remarkably, this core value, which applies to society as a whole, is often not translated to the mutual relationships between people in organizations. As you read in Chapter 3, both the union and the works council were set up to put employees on a more equal footing with the top of the organization. But shared control can go a step further. In the previous section, you learned about shared ownership. The additional effect of this is that an employee-shareholder gains voting rights at the general meeting of shareholders and automatically shares in the profits. This changes their situation from a subordinate to an equal position. An equal position in decision-making can come with or without equity participation. An example of a form that regulates control without share participation is sociocracy. Chapter 5 shows what this looks like in practice. Variants are self-managing and self-organizing teams and democratic organizations. In all variants, there is no top-down manager telling them what, where, when and how to produce and what happens to the profits.

People Are Motivated to Be of Value

In Chapter 3, you read about motivation as a means of production. You learned that this also fits with the theory about the market: if you work for someone else who gets to make decisions about you, and who pockets the profits, what is your interest in working hard?

> A son, who works part-time at a snack bar to earn additional income, shares an exciting idea with his mother: "I have a great idea to boost our sales during Christmas. We could create special Christmas boxes for children with a surprise inside. However, I hesitate to propose it because it would mean extra work for me without any additional benefits."

What about motivation when organizations are set up on the basis of equality, in which control and the added value produced by everyone is shared? As mentioned earlier in Chapter 3, Gagné and Deci (2005) argue that people are self- or autonomous motivated when three basic psychological needs are met: autonomy, competence and connectedness. They developed a theory that assumes that people should be able *to decide for themselves* (self-determination), that they should be able to (co-)decide on matters that concern them. We do this every day in our private lives. We decide for ourselves what clothes we want to wear, how much money we spend and what we eat at night. However, if you have roommates, you often agree on what someone likes or dislikes or has special dietary requirements. You also need to *have the skills to get* the job done. Little is as frustrating as doing a maths problem you don't understand. Building a house when you can't even hold a hammer leads to failure. Motivation also increases when you understand that your *contribution aligns with the larger purpose of meaningful work*. Translated to organizations, this means that you have a say in decisions that affect you and your team, you have opportunities to develop your skills and competencies, and you feel a sense of connection to the broader purpose and goals of the organization. Sociocracy and democracy are based on precisely these three needs (Georges & Romme, 1999).

Free and equal in dignity, endowed with reason and conscience and a spirit of brotherhood, are different principles than those on which our current society is built. If we as a society want to move to a sustainable economy, companies will no longer employ people as unmotivated production tools, but as equal production partners (as brothers and sisters) who want to contribute of their own accord.

4.3 Multiple Growth for All

In Chapter 3, you learned that it is generally accepted that a market that operates as freely as possible is the best recipe for economic growth. However, not everyone holds this view. Hans Achterhuis, philosopher and writer, states in his book *The Utopia of the Free Market* that economic growth was greater when the market was subject to all sorts of restrictions. And Arjan Broers explains in his book *Geld en Goed (Money and Good)* that growth is mainly caused by higher prices and is based on financial, ecological and social debt.

You have also read in Chapter 2 that our way of production and related economic growth have many harmful side effects. Robert Kennedy, an American politician and brother of the American president John F. Kennedy, also saw this. He gave a speech just before his own death, where he talked about the negative effects of economic growth:

How GDP failed. Robert Kennedy. University of Kansas, 18 March 1968[18]

Too much and for too long, we seemed to have surrendered personal excellence and community values in the mere accumulation of material things. Our Gross National Product, now, is over $800 billion dollars a year, but that Gross National Product—if we judge the United States of America by that—that Gross National Product counts air pollution and cigarette advertising, and ambulances to clear our highways of carnage.

It counts special locks for our doors and the jails for the people who break them. It counts the destruction of the redwood and the loss of our natural wonder in chaotic sprawl.

It counts napalm and counts nuclear warheads and armoured cars for the police to fight the riots in our cities. It counts Whitman's rifle and Speck's knife, and the television programs which glorify violence in order to sell toys to our children.

Yet the gross national product does not allow for the health of our children, the quality of their education or the joy of their play. It does not include the beauty of our poetry or the strength of our marriages, the intelligence of our public debate or the integrity of our public officials.

It measures neither our wit nor our courage, neither our wisdom nor our learning, neither our compassion nor our devotion to our country, it measures everything in short, except that which makes life worthwhile.

In the previous sections, you learned about seven guiding principles for a sustainable market economy. Those guiding principles together lead to sustainable growth, the eighth guiding principle for a sustainable market (Fig. 4.5).

But what exactly is sustainable growth? And do sustainability and growth go hand in hand? Back to economic growth: what exactly is economic growth? It means adding more value to a product than that product had before. But does this mean that it has a direct negative effect on our planet and ourselves? Suppose you decide to start a company where old mobile phones are dismantled and you sell the parts. It seems odd, but with this business you are actually adding value: you reusing what otherwise would have been waste. Or say you collect old carpets, strip them down to the bone

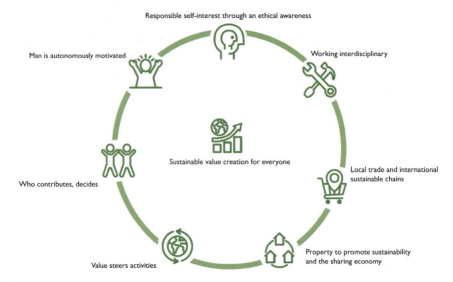

Fig. 4.5 Eight guiding principles of a sustainable market economy

and turn them into new yarn, which is then bought by furniture makers. The energy you use in your factory is one hundred per cent sustainable. In that case, your company has also contributed to economic growth, but in a sustainable way. So economic growth does not necessarily lead to environmental damage or bad effects for people. Growth does not automatically mean that we burden the earth more.

Daniël Mügge, professor at the University of Amsterdam and correspondent at *Follow the Money*, also shares this view. He argues that 'even if you only look at this narrow definition of economic growth, it goes very well with a transition to sustainability. That transition requires huge investments. If they are made, they translate directly into extra jobs and a higher Gross Domestic Product'.[19] Sustainability can therefore go hand in hand with economic growth. Mügge does argue, however, that the investments required for sustainable production must be paid for. Who wants to pay for my recycled raw materials if they can be extracted much more cheaply in Congo? If you don't see a market, and neither do investors, then this business is unlikely to have a long life. The compelling hand of the market—if you are not better in price, quality and innovation than your competitor you go out of business—discourages sustainable start-ups and prevents many existing companies from making the transition to a sustainable production process.

So if economic growth is linked to a production process that involves a lot of harm, then the link between prosperity and economic growth is less and less appropriate. But imagine a country where value is added in a fully sustainable way, and where the products actually make our lives better. Is measuring economic growth then still too limited an indicator?

Below is a brief explanation of both points.

A broad welfare monitor

A broader perspective can reveal that economic growth can conflict with, for example, people's health, happiness and income equality. One of the first countries to calculate prosperity more broadly than economic growth alone was Bhutan. In this country, the gross national happiness was invented as an alternative to gross national product. In 2009, the OECD took the initiative to supplement figures on economic growth with figures on people's well-being. This resulted in '*The OECD Framework for Measuring Well-Being and Progress*'[20] (Fig. 4.6).

Fig. 4.6 The OECD dashboard to map well-being in a given year and future well-being (*Source* OECD)

The monitor measures prosperity using more indicators than GDP alone. Increased prosperity then becomes an increase in value. For companies, this means a shift from profit to value maximization.

Limiting growth

The question of whether we should continue pursuing the concept of growth remains a subject of debate. One group emphasizes this question and promotes the idea of a steady state economy. In essence, this approach advocates for not exceeding the Earth's carrying capacity. On the other hand, there is a more radical group that argues that only through degrowth can we achieve a sustainable planet. This perspective suggests a deliberate reduction in economic activity and consumption to alleviate environmental pressures: 'Degrowth proponents advocate reducing ecologically destructive forms of production and resource throughput in wealthy economies to achieve environmental goals, while transforming production to focus on human well-being' (Bodirsky et al., 2022). How humans can transform unlimited growth into multiple growth for all is described in Chapter 5.

4.4 The Role of Government in a Sustainable Market

Finally, the government plays a vital role in promoting collective sustainability as it is responsible for creating and implementing agreements that apply to society as a whole. For instance, in the Netherlands, there has been a significant decline in the population of meadow birds in recent years. Agriculture, particularly the pressure on farmers to increase production for a competitive milk price, is a major contributing factor affecting the habitat quality for these birds. While individuals may choose to purchase more expensive milk to support bird conservation, the effectiveness of such actions relies on widespread consumer participation. On the other hand, a milk producer can opt to adopt a fair pricing strategy, but if their competitors continue to offer lower-priced alternatives and consumers remain price-sensitive, the sustainable producer may struggle to remain competitive. In such cases, government intervention can be crucial. One approach is to impose a tax on milk prices and allocate the revenue as subsidies to farmers, ensuring a fair and sustainable income for their efforts (Image 4.4).

The extent to which the government intervenes to promote sustainability depends on which parties are in power. In the Netherlands, the government is

Image 4.4 A fair price for milk can increase biodiversity (*Source* Pixabay, CallyL)

implementing measures to establish favorable conditions and impose restrictions in order to foster sustainability and mitigate the adverse impacts of the free market. A notable example of a measure aimed at creating favorable conditions is the provision of numerous subsidies for collaborations between companies and educational institutions like universities and colleges. These collaborations focus on developing sustainable solutions, such as clean energy and sustainable labor practices. Additionally, subsidies are also offered for the purchase of electric cars, aligning with the government's commitment to promoting environmentally friendly transportation options. Furthermore, the government has also implemented restrictive measures. One prominent example occurred in 2012 when the European Union introduced a ban on battery cages for chickens, following the revelation of the deplorable conditions in which these animals were kept. And as of 1 October 2019, everyone is obliged to fill up with cleaner petrol, the so-called E10. The EU's proposal to introduce a price cap on the price of gas (September 2022) is another example of a restrictive measure.

The degree to which the government imposes restrictions depends in part on public support. This was clearly visible during the corona pandemic. After the alarm bells went off in China and Italy, restrictive measures were quickly taken to prevent further rapid spread of the virus. The images of overcrowded ICU wards and the acute danger of infection and possible death ensured great support for these measures. The catering industry had to close down and people were obliged to keep one and a half metres distance from each other. In June 2020, the number of hospital admissions and ICU occupancy rates had fallen so far that a number of restrictions could be lifted. The measures that remained (such as the use of masks on public transport and keeping a one and a half metre distance from each other) provoked resistance from some people: support declined as the risk of infection and death decreased. Protests arose with police intervention, and people wondered whether the restrictions were leaning too much towards a police state or dictatorship.

The pandemic has highlighted the importance of an immediate and widespread high risk that affects everyone, which has helped in garnering support for restrictive measures. However, when it comes to sustainability, such an acute risk is often absent for many people. The risks associated with sustainability issues may be perceived as affecting others or being distant in the future. This poses a challenge in gaining support for restrictive measures aimed at mitigating temperature increase and addressing long-term sustainability concerns.

Global policy

Many measures to promote sustainability require a European and global approach. An example of this is solving the climate issue. After all, CO_2 emissions do not stop at national borders; climate change is a global phenomenon. Every country will have to take steps in this direction.

One step taken at European level in 2005 was the introduction of the EU Emissions Trading System (EU ETS).[21] This system imposed limits on CO_2 emissions on more than 11,000 heavy energy-consuming installations (power plants and industrial plants) and airlines operating between these countries. The cap on emissions decreases each year. If a participant exceeds the limit, in principle a fine is imposed. But there is also a market for the right to emit a certain amount: if you emit less than what is allowed, you can offer the surplus on the market to companies that would like to emit more than their share. With this system, by 2020 21% less CO_2 will have been emitted than on the reference date of 2005, and according to the EU that is on schedule.

> **Example**
> (Richard) **Sandor: The free market is the weapon against pollution**
> The idea of a market for pollution was not new. For example, in the eighties of the former century, acid rain caused respiratory and skin problems and building damage. In 1990, the government passed the Clean Air Act, which determined how much sulphur manufacturers could emit. Companies could invest in cleaner factories, but they could also buy pollution rights from others who had cleaned up more than was required. The price of the right to emit

> a ton of sulphur was 300 dollars at the first trade. Now that has dropped to $107. Acid rain is no longer a problem, which shows the market is working.[22]
> Source: *Trouw* (Dutch newspaper), 14 March 1998

The global awareness that climate change is a serious threat to human security has also resulted in climate agreements that governments worldwide reach with each other during United Nations climate conferences (the COP: Conference of the Parties). The first climate treaty was concluded in 1992 in Rio de Janeiro, Brazil, and came into force in 1994. Since then, a conference has been held every year to discuss the implementation of the agreements and to negotiate an adaptation of the climate treaty.[23] At the 2015 conference (the COP21, also known as the Paris Agreement), it was decided that all countries should set out their efforts to reduce CO_2 emissions in national plans, with the aim of keeping global warming below 2 degrees Celsius.

> During the COP21 in Paris, governments came to the following agreements on reducing emissions:
> - 'a long-term goal of keeping the increase in global average temperature to well below 2 °C above pre-industrial levels;
> - to aim to limit the increase to 1.5 °C, since this would significantly reduce risks and the impacts of climate change;
> - on the need for global emissions to peak as soon as possible, recognising that this will take longer for developing countries;
> - to undertake rapid reductions thereafter in accordance with the best available science, so as to achieve a balance between emissions and removals in the second half of the century'.[24]

But sustainability is broader than just the climate. Other environmental agreements made at European level include, for example, a ban on the use of neonicotinoids (a pesticide) after it became known that it had a direct effect on the bee population. Examples of the European Union's policy themes in the social field are combating poverty, funding projects to make production chains more sustainable[25] and protecting and improving health.

> At the European level we are committed to a sustainable market for chemicals and their use. Regulations[26] on this subject are part of the legislative system REACH. Their website says the following:
>
>> REACH is a regulation of the European Union, adopted to improve the protection of human health and the environment from the risks that can be posed by chemicals, while enhancing the competitiveness of the EU chemicals industry. It also promotes alternative methods for the hazard assessment of substances in order to reduce the number of tests on animals.
>>
>> In principle, REACH applies to all chemical substances; not only those used in industrial processes but also in our day-to-day lives, for example in cleaning products, paints as well as in articles such as clothes, furniture and electrical appliances. Therefore, the regulation has an impact on most companies across the EU.[27]

4.5 Summary

The guiding principles of the transition to a sustainable market economy

The free market economy operates based on eight guiding principles that, in conjunction with our behavior, shape its functioning. A transition to a sustainable market economy requires different behaviour and new guiding principles.

The eight guiding principles of a sustainable market economy

The sustainable market economy has the following guiding principles:

1. *Responsible self-interest through ethical awareness.* In a sustainable market economy we take our responsibility for the earth and other human beings and accept that our freedom has limits. Sustainability is a moral shell around self-interest and the compelling hand of the market and can be reasoned from three ethical perspectives:
   - Utilitarianism. According to Bentham (1748–1832) and Mill (1806–1873), people who think in terms of utility believe that you behave well if your behaviour leads to increased happiness for as many people

as possible rather than harming people. Said differently: you do well if you deliver the greatest good for the greatest number.
- Duty. Maximizing utility sometimes leads to the wrong decisions. The philosopher Kant (1724–1804) argued that there are universal values that transcend all utility, such as the right to life, honesty and justice. These values dictate our duties.
- Virtue ethics. Virtue ethics is about *wanting to be* a good person. Aristotle (384–322 BC) already laid the foundation for virtue ethics. He argued that people are inclined to goodness and are most happy when they develop certain virtues.

Sustainability as a theme particularly emerged after the publication of the report *Limits to growth* by the Club of Rome. This club was founded in 1968 and consisted of a group of scientists and other influential people who were concerned about the depletion of our earth. Since then, the theme has received more and more attention worldwide.

2. *Interdisciplinary work for sustainable solutions.* Assuming responsibility becomes feasible when one has the ability to influence the outcome. However, due to the nature of work specialization, individuals are often limited to making contributions within their specific sub-domains. To address this limitation and effectively tackle *wicked problems*, interdisciplinary collaboration is essential.
3. *Local trade and international sustainable chains. Firstly,* countries can impose import levies or quotas on products from abroad. Secondly, companies can make the international production chain more sustainable. Finally, you learnt about the positive effects of local production and consumption.
4. *Ownership to promote sustainability and the sharing economy.* Wider ownership gives more people control over production, and thus the opportunity to make it more sustainable. We could also get rid of ownership entirely and/or increase the sharing economy.
5. *Multiple value drives activities.* In a sustainable economy, consumers look at the sustainable value of the product in addition to its cost. They pay a fair price and make investments with a sustainable impact.
6. *Who contributes, decides.* Equality is fundamental to a sustainable economy as it places the value of human life as paramount and works towards reducing financial inequality, consequently addressing the growing issue of social inequality. Equality entails recognizing the intrinsic worth of individuals and rejecting their commodification for the sake of cost reduction

or heightened productivity. In both organizational and societal contexts, equality encompasses equitable participation in decision-making processes and a fair distribution of profits.
7. *Man is autonomously motivated.* Once the position of individuals within organizations shifts from subordination to equality, a fundamental human need is met: the right to self-determination. This represents one of the three human needs that positively impact autonomous motivation. The other two needs are having the necessary skills to accomplish tasks and perceiving one's contribution as meaningful within the larger context of work.
8. *Multiple growth for all.* Economic growth means adding more value to products than the year before. In a country where fully sustainable value is added and where products truly make our lives better, economic growth and sustainability go hand in hand. But we are not there yet. A broader perspective can reveal that economic growth can conflict with, for example, people's health, happiness and income equality.

The role of government in the sustainable market economy

Many countries have mixed economies. The extent to which the government stimulates or regulates depends on the political parties in power. The global awareness that climate change is a serious threat to human security has resulted in climate agreements that governments worldwide make with each other during United Nations climate conferences.

Continue reading and watching

- Kate Raworth's donut model is an appealing way to gain more insight into the idea of sustainability. Her website is full of information.
- Want to delve further into ethics? Then take watch the *Justice* lecture series by Michael Sandel. In 24 one-hour lectures he covers a wide range of ethical theories and examples.
- If you want to learn more about equality, listen to Richard Wolff's one-hour podcast on Marx's theory.
- A TED talk by Richard Wilkinson will give you more insight into the effects of inequality between people in society.

Notes

1. Would you like to study Adam Smith and his work? I used a few interesting sites and videos for this book, such as: www.adamsmith.org/the-theory-of-moral-sentiments and this documentary about him: https://youtu.be/V6S6pMsKzlI. In fact, his own work is quite difficult to read, but can be found here: https://ibiblio.org/ml/libri/s/SmithA_MoralSentiments_p.pdf (accessed October 2, 2022).
2. This example is dealt with, among others, in the lecture series on ethics mentioned in Chapter 3 by Michael Sandel. His lecture with this example can be found here: http://justiceharvard.org/lecture-2-the-case-for-cannibalism/#1477504398584-5d76ef29-aab0 (accessed October 2, 2022).
3. Thinking big, moving small. Counterpoints in sustainable decision-making. Accessed September 17, 2022, from http://fluvere.nl/documenten/Grootdenken.pdf.
4. What on Earth is the doughnut?... Accessed October 2, 2022, from www.kateraworth.com/doughnut/.
5. Promote sustainable development. Accessed October 2, 2022, from https://www.un.org/en/our-work/support-sustainable-development-and-climate-action.
6. The Intergovernmental Panel on Climate Change. Accessed October 2, 2022, from www.ipcc.ch/.
7. Take action for the Sustainable Development Goals. Accessed October 2, 2022, from www.un.org/sustainabledevelopment/sustainable-development-goals/.
8. The Earth Charter. Accessed October 2, 2022 from https://earthcharter.org/.
9. The Urgenda climate case against the Dutch government. Accessed October 2, 2022, from https://www.urgenda.nl/en/themas/climate-case/.
10. The wicked problems plaza. Accessed October 2, 2022, from https://repub.eur.nl/pub/93223.
11. ETI, what we do. Accessed September 20, 2022, from https://www.ethicaltrade.org/about-eti/what-we-do.
12. The image that the sharing economy is still at the neighbourhood level, where residents use a sharing platform to lend each other a sander or prepare meals, needs to be adjusted: P+ SPECIAL: Big companies are conquering the sharing economy. Accessed October 4, 2020, from www.p-plus.nl/nl/nieuws/Grote-bedrijven-deeleconomie.
13. Read the entire critique here: www.vn.nl/liefdeloosheid-in-kapitalistische-tijden/ (accessed October 2, 2022).
14. Sleipnir Foundation. Accessed October 2, 2022, from www.stichtingsleipnir.nl/.
15. Annual report PGGM 2021. Consulted on 2 October 2022, from https://www.pggm.nl/media/nuep1jz1/annual-report-of-pggm-n-v-2021.pdf.

16. Oikoicredit working method. Accessed October 2, 2022, from https://www.oikocredit.coop/en/what-we-do/what-we-do.
17. A history of microfinance | Muhammad Yunus | TEDxVienna Accessed September 22, 2022, from https://youtu.be/6UCuWxWiMaQ.
18. Bobby Kennedy on GDP: 'Measures everything except that which is worthwhile'. Accessed October 2, 2022, from www.theguardian.com/news/datablog/2012/may/24/robert-kennedy-gdp.
19. Sustainability is economic growth? Accessed October 2, 2022, from www.ftm.nl/artikelen/duurzaamheid-is-economische-groei?share=1.
20. Measuring well-being and progress: Well-being research. Accessed October 2, 2022, from www.oecd.org/statistics/measuring-well-being-and-progress.htm.
21. EU Emissions Trading System (EU ETS). Accessed October 2, 2022, from https://climate.ec.europa.eu/eu-action/eu-emissions-trading-system-eu-ets_en.
22. Sandor: The free market is the weapon against pollution. Accessed October 2, 2022, from www.trouw.nl/nieuws/sandor-de-vrije-markt-is-het-wapen-tegen-vervuiling-b5bdff68/.
23. You can read more about the background of the climate treaty on the page of the initiator of this treaty, the United Nations (UNFCCC = United Nations Framework Convention on Climate Change). Accessed October 2, 2022, from https://unfccc.int/.
24. Paris Agreement. Accessed September 23, 2022, from https://climate.ec.europa.eu/eu-action/international-action-climate-change/climate-negotiations/paris-agreement_en.
25. Eliminating child labour and forced labour in the cotton, textile and garment value chains. Accessed October 2, 2022, from www.ilo.org/ipec/projects/global/WCMS_649126/lang--en/index.htm.
26. Regulations contain rules that are directly applicable in all member states of the European Union. This is called 'direct effect'. Regulations therefore have a similar status to national laws in the member states, but in the event of a conflict, the regulation takes precedence over national law. For more information, see: https://ec.europa.eu/info/law/law-making-process/types-eu-law_en (consulted on 2 October 2022).
27. Understanding REACH. Accessed October 2, 2022, from https://echa.europa.eu/regulations/reach/understanding-reach.

Literature

Adams, R. B., Licht, A. N., & Sagiv, L. (2009). *Shareholderism: Board members' values and the shareholder-stakeholder dilemma*. CELS 2009 4th Annual Conference on Empirical Legal Studies Paper.

Blasi, J., Kruse, D., & Freeman, R. B. (2018). Broad-based employee stock ownership and profit sharing: History, evidence, and policy implications. *Journal of Participation and Employee Ownership, 1*(1), 38–60.

Bodirsky, B. L., Chen, D. M.-C., Weindl, I., Soergel, B., Beier, F., Molina Bacca, E. J., Gaupp, F., Popp, A., & Lotze-Campen, H. (2022). Integrating degrowth and efficiency perspectives enables an emission-neutral food system by 2100. *Nature Food, 3*(5), 341–348.

Elkington, J. (1999). *Cannibals with forks: The triple bottom line of 21st century business*. Wiley.

Gagné, M., & Deci, E. L. (2005). Self-determination theory and work motivation. *Journal of Organizational Behavior, 26*(4), 331.

Georges, A., & Romme, L. (1999). Domination, self-determination and circular organizing. *Organization Studies, 20*(5), 801–832.

Zander, B., & Zander, R. S. (2002). *The art of possibility*. Penguin Books.

5

The Happiness of a Sustainable Market

The adoption of these new guiding principles paves the way for the emergence of a sustainable market. In a sustainable market economy, we embrace our responsibility towards others and nature, strive for value maximization, and promote equality among individuals. The foundation of a sustainable market is a livable Earth and the well-being of humanity. Despite these considerations, a sustainable market still operates on the fundamental principles of supply and demand, recognizing the inherent value and effectiveness of the market as a remarkable human innovation. What does a sustainable market look like and what strategies do companies apply in a sustainable market? This chapter will address these questions.

5.1 How the Sustainable Market Works

A sustainable market, although not fully realized yet, encompasses several key components that are gradually emerging. In envisioning the ideal sustainable market, let us return to the market square described in Chapter 1. In this scenario, merchants arrive at the square using carbon-neutral vehicles, and their stalls showcase sustainably produced goods. This implies that production processes prioritize the well-being of the planet and its inhabitants. For instance, market vendors ensure they pay fair prices to their suppliers, and profits and wages are distributed equitably within their organizations. On the consumer side, there is a growing demand for sustainable products at fair

© The Author(s), under exclusive license to Springer Nature
Switzerland AG 2023
M. Boersma-de Jong and G. de Jong, *Foundations of a Sustainable Market Economy*, https://doi.org/10.1007/978-3-031-28186-0_5

prices. Consumers are willing to support this because they understand that producers are not solely driven by profit and that the product offers multiple values. The price reflects the true costs incurred, and the fair distribution of wages ensures sufficient purchasing power among the population to afford these slightly more expensive but sustainable products.

In the sustainable market, the pursuit of self-interest remains a fundamental aspect. For instance, individuals may continue to engage in baking bread for the entire village because it offers them an opportunity to earn money. Similarly, as consumers, we still make purchases based on our essential needs for survival or convenience. Thus, the market continues to be driven by the mutual interests of both producers and consumers. In terms of company strategies, they remain largely unchanged. In the old economy, strategies such as product innovation, quality, customer relationships, and competitive pricing were crucial for survival and profitability. These strategies were effective because they aligned with consumers' primary selection criteria. In the sustainable market, these strategies still apply, with the key difference being that all available products are inherently sustainable. In this context, there is no longer a trade-off between cheap but unsustainable options and more expensive yet sustainable alternatives. We no longer have to choose between a polluting iPhone and a sustainable Fairphone, or between clothing made by exploited child labor in India and ethically produced, sustainable clothing by adults who receive fair wages. Brand choices remain, but all brands in the sustainable market prioritize sustainability.

In the future sustainable market, companies have a broader purpose beyond serving the interests of shareholders alone. Instead, they prioritize the needs and well-being of all stakeholders involved. The focus shifts from profit maximization to value maximization, where companies strive to create value for society, the environment, employees, customers, and other stakeholders. Shareholders in this market invest in companies that embrace and deliver multiple forms of value, recognizing the interconnectedness of economic, social, and environmental aspects. In a sustainable economy, banks also look at the multiple value of companies. Sustainable banks, such as Triodos Bank and Grameen Bank, will be the example for traditional banks in the future. They will only offer loans to companies that create multiple value. The strategy of banks aligns with the principles of sustainability. It is inconceivable that banks would prioritize their own financial gains by selling products to customers without considering the risks involved. Instead, banks adopt a responsible approach where the well-being and interests of customers are paramount. They provide transparent information and ensure that customers have a comprehensive understanding of the products and services offered.

5 The Happiness of a Sustainable Market

In the sustainable economy, consumers have transformed into conscious buyers who prioritize mindful consumption. They opt for buying less overall and prioritize products that are recycled or have minimal environmental impact. In this paradigm, the concept of ownership no longer defines their identity as much as the positive impact they create for others and the world around them. The emphasis shifts towards the meaning and value they contribute to society, fostering a sense of purpose and connection.

Finally, the government plays a crucial role in fostering a sustainable market economy by establishing the necessary framework for its operation. Guided by sustainable principles, the government formulates and oversees regulations that set the conditions and restrictions for sustainable production. The underlying ideology of the government in a sustainable market economy is rooted in the belief that through cooperation, consumers can ultimately access the most sustainable products at fair prices. This approach aims to enhance the well-being of humanity, the planet, and all living beings, promoting a harmonious balance between economic prosperity and ecological stewardship.

During the 2019 nitrogen crisis, a notable example of a restrictive measure that was implemented to benefit the environment was the implementation of nitrogen reduction policies. 'Nature and environmental organizations had initiated the cases because they believed it was unfair for the Dutch government to assume a decrease in nitrogen deposition in the Netherlands and allow companies to expand accordingly. Due to the large amount of nitrogen precipitation, the biodiversity in vulnerable nature areas in the Netherlands has been under pressure for years. Plant species are declining in number, causing birds and insects to disappear from the areas'.[1] As a result, the Council of State invalidated the Programma Aanpak Stikstof (PAS) in May 2019. This decision led to the halt of numerous construction projects and livestock expansions. To alleviate the significant construction-related disruptions, the leading liberal party had to implement an unprecedented measure: reducing the speed limit on roads from 130 to 100 km per hour. However, this measure also affected farmers, which subsequently triggered the farmers' protests. These protests involved tractors on the motorways and demonstrations on the Malieveld in The Hague (Image 5.1).

Image 5.1 Small-scale farming has declined sharply (*Source* Pixabay, analogicus)

The contrast between our current economy and the ideals of sustainability is striking. However, within the sustainable economy framework, there is hope for positive change. By adhering to sustainable guiding principles, we can begin to address and eliminate the negative ecological and social externalities associated with our current economic system. While we have not fully reached our destination, signs of transition towards sustainability are becoming increasingly apparent. More and more companies are emerging with a strong commitment to contribute to a sustainable world. There are many rankings, national and international, where companies are only too happy to be listed. Just search via Ecosia (an alternative sustainable search engine for Google) for 'top sustainable companies' (each list often leads to discussions about the criteria) to get an impression. Sometimes companies appear to be very sustainable on paper, but in practice are much less so. This is called *greenwashing*.

In this chapter, you will learn how companies shape sustainability. There are three main sections:

1. a sustainable mission and vision, aimed at creating value for people and the environment (Sect. 5.2);

Fig. 5.1 Companies can operationalize sustainability in three ways

2. circular and value-driven product innovation (Sect. 5.3);
3. multiple cost reductions (Sect. 5.4) (Fig. 5.1).

The foundation of many sustainable entrepreneurs lies in their personal commitment to living a sustainable life. This commitment is often evident in the mission and vision of sustainable businesses.

5.2 Translating Guiding Principles into a Sustainable Mission and Vision

A mission or vision is considered sustainable when it focuses on improving the environment and human well-being. Sustainable companies aim to generate a positive impact that aligns with these principles. The Body Shop is an exemplary company that embraced sustainable guiding principles early on in its mission and vision. Founded in the 1970s by Anita Roddick, the Body Shop made a deliberate choice to sell only cosmetics and beauty products that were not tested on animals. This decision demonstrated the belief that business and doing good can be interconnected. The company's vision for a fairer and more beautiful world resonated with consumers. By promoting and selling their products, the Body Shop also raised awareness about the issue of animal testing. This sustainable strategy not only created change within the company but also inspired a broader shift in consumer attitudes. The increased awareness surrounding the use of animals in cosmetics ultimately led to a ban on animal testing in Europe since 2013. However, it is important to note that imported products from countries like China may still undergo animal testing.

> **Tip**
> It is now known that the cosmetics of the Body Shop are animal free. But many more brands have made this change. Do you want to know if your cosmetics are free of animal testing? Go to the website of *Shop like you give a damn* and find out if your cosmetics are on the list.

Another great example of a sustainable mission is that of Tony's Chocolonely: 'together we make 100% slavery-free the norm in chocolate'. The story of this company starts with Teun van de Keuken, who immersed himself in the production of chocolate for the Dutch TV program Keuringsdienst van Waarde and was shocked to discover that (child) slavery still exists. The company has since made a big impact in the cocoa industry (Image 5.2).

The company only buys cocoa from its partner cooperatives, which are visited regularly. They also distribute extra profit to the suppliers and invest in improving the production process. With their company they not only make delicious chocolate, but they also improve the living conditions of people. They have also made western consumers aware of the damage the production of chocolate can cause.

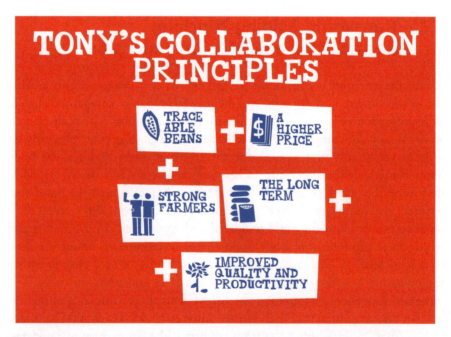

Image 5.2 The impact of Tony's Chocolonely (*Source* Tony's Chocolonely)

5 The Happiness of a Sustainable Market

Here is a list of some inspiring companies:

- Fairphone: '*From the earth to your pocket, a smartphone's journey is filled with unfair practices. We believe a fairer electronics industry is possible. By making change from the inside, we're giving a voice to people who care*'.
- Triodos Bank: '*Money has the power to change. Every action, however small, sets something in motion. Together we can make small the new big*'.
- Patagonia: '*We know that our business activity-from lighting stores to dyeing shirts-is part of the problem. We work steadily to change our business practices and share what we've learned. But we recognize that this is not enough. We seek not only to do less harm, but more good*'.
- Mondragon corporation: '*Humanity at work.*'
- Tesla: '*Tesla's mission is to accelerate the world's transition to sustainable energy*'.
- De Nieuwe Band: '*Together with others, we are actively working towards a world of conscious consumption and production of social, sustainable, tasty products*'.

A company's mission can be a source of inspiration for people to take steps towards a sustainable and virtuous life. With your company you still act from your own interest (you offer products on the market to make money), but you also serve the interest of others and nature—not at the expense of, but in harmony with.

> Entrepreneurs who have embraced a sustainable mission and vision often do so driven by personal motivations. They have transitioned from focusing solely on the "what" and "how" of their business to living a life guided by the "why." When living from the "what" and "how," conversations at social gatherings may revolve around their profession and the financial gains they can achieve. However, when living from the "why," these entrepreneurs are more inclined to discuss the underlying reasons and purpose behind their actions. Sustainable entrepreneurs often want to live a meaningful life (you read about it earlier in Chapter 4), in which they deliver value for themselves and society. Simon Sinek explains living from the why in his famous video *Start with why* (Fig. 5.2).

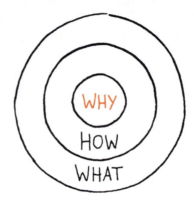

Fig. 5.2 Sinek's golden circle

Sustainable companies translate their mission and vision into concrete sustainable products and production processes. How they do this is explained in the following sections. The classification largely follows that of Chapters 1 and 2.

5.3 Sustainable Product Innovation and Improvement

From Linear to Circular

In Chapter 2, you learned that a continuous flow of new products, driven by factors such as product innovation, improvement, quality, and customer relationships, can have associated negative impacts. Overexploitation, climate change and a plastic soup are just a few of the negative externalities. But what if we apply the new guiding principles outlined in Chapter 4? Can product innovation and improvement then lead to a much more sustainable economy? The answer is a resounding yes!

They called their idea *cradle-to-cradle (C2C)*: what is born, eventually finds itself again in the birth of something else. In a *Tegenlicht documentary* from 2007,[2] McDonough and Braungart give an example of such a product innovation: develop a foil for ice creams that is strong when frozen, and turns into water when thawed. Mix this foil with wild seeds, so that each foil can be thrown away in nature. Waste is then the basis for a proliferation of new flowers and plants. The appeal of their idea is that we are allowed to live exuberantly, after all, nature is not limited, but grows and blossoms, and produces enormous amounts of waste every autumn. Death is the breeding ground for new life. So can products be designed in such a way that they can always serve as a breeding ground for new products? The basis must lie in designing material that is organic and can therefore be absorbed by nature. If this is not possible, reuse the material. This could be done through *upcycling* or *downcycling*. With upcycling, existing material is processed into a new product with more value. A good example is a suitcase whose exterior is made from recycled PET bottles. Downcycling is the opposite: the material is reused, but with a lower value and reuse becomes more difficult after each application.

Making products and production processes sustainable (you will learn more about the latter in the next section) through product innovation and improvement falls under the heading of the *circular economy*. Remember the linear production process from Chapter 2, where there were three elements: extraction, consumption and waste? In a circular economy, we ensure that raw materials do not reach the stage of waste and reduce our ecological footprint. The Netherlands Environmental Assessment Agency (PBL) presents the various circularity strategies in Image 5.3.

Table 5.1 provides information on exactly what each strategy entails.

The first strategy, *refuse*, is the most effective for a sustainable market: stop producing non-sustainable products and possibly replace them with a sustainable variant. Do we really need mermaid tails that we can buy cheaply from a webshop? Or can we do without them?

Do you need a replacement? Then *rethink* is the second strategy. McDonough and Braungart's ice cream film is a good example. Replacing fossil raw materials with raw materials that are part of nature is a '*biobased*' alternative. In this quest for circularity, we can also learn a lot from how nature itself arrives at solutions. This is called *biomimicry* and involves studying how nature comes up with solutions and applying what we learn to innovations for people.

> **Tip**
>
> If you want to know more about *biomimicry*, go to the website of Janine Benyus, who published the book *Biomimicry, Innovation Inspired by Nature* in 1997.

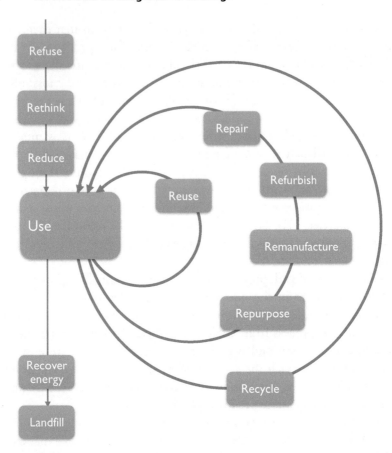

Image 5.3 Methods for circular living and business (*Source* PlanBureau voor de Leefomgeving (PBL))

If a replacement product cannot be developed, investigate whether you can reduce the use of raw materials by using products more intensively and producing them more efficiently. These three strategies, *reuse*, *repair* and *refurbish*, are all aimed at introducing sustainability into product design. As for the many products that have not been designed sustainably, we should use them as long as possible (*reuse*), and if a product breaks down, hand it in for repair or improvement (*refurbish*). In these three strategies, the original product remains intact. If the product has too many defects, parts can be reused in the same product (*remanufacture*) or in other products (*repurpose* and *recycle*). Finally, the remains of the product form a possible basis for energy *recovery*.

Table 5.1 The different strategies with explanations

Refuse	Make product redundant by renouncing its function, or replace it with a radically different product
Rethink	Intensify product use (e.g., by sharing products or multifunctional products)
Reduce	Make product more efficient to use or manufacture by reducing the amount of raw and auxiliary materials in the product
Reuse	Reuse of discarded product in the same function by another user
Repair	Repair and maintenance of defective product for use in its previous function
Refurbish	Refurbishment and modernization of old product for use in improved version of its former function
Remanufacture	Use parts from discarded product in new product with the same function
Repurpose	Use discarded product or parts thereof in new product with different function
Recycle	Processing materials to the same (high-grade) or lesser (low-grade) quality
Recover energy	Incineration of materials with energy recovery

Note The darker green, the more the product contributes to a circular economy (Color figure online)

Below are some practical examples, both from home and business situations.

Circularity @Home

Some strategies are centuries old and were used early on at home. For example, old bread is still delicious if turned it into French toast: bread dipped in a mixture of milk, egg and cinnamon, fried with a little butter in the frying pan. It's a great example of the *rethink strategy*. Instead of using an extra bleaching detergent, white laundry was often put outside after washing: on the grass and with the sun, all stains disappeared. Socks were darned, shoes patched many times and the crib survived many generations of babies. Maybe you had to wear your older brother or sister's clothes when you were little? If so, you were sustainable early on! Many of these strategies came from poverty—there wasn't enough money to buy anything new. Even now, plenty of families have to choose second-hand items (reuse) not from the ideal of recycling, but from a lack of money. Alternative materials are now being sought for clothing. Still, every fabric has its advantages and disadvantages.

Business strategies

In business terms, it is increasingly possible to earn a good living from circularity. The most obvious transition in recent years is the development of wind and solar energy. This energy source is an example of the refuse strategy: polluting energy sources are being replaced by sustainable sources. Whereas until recently we were dependent on oil and gas and on large companies exploiting them, nowadays every individual can instal a number of solar panels and thus become their own energy supplier. The possibility of generating energy locally has led to the creation of energy cooperatives: individuals who invest in the development of energy generation by, for example, jointly purchasing a windmill. The proceeds of energy projects can be returned to the investors or reinvested in sustainable projects. While solar panels can be seen as an example of the refuse strategy, issues remain regarding the non-sustainable production of these panels.

Residents of Reduzum (The Netherlands) were early adopters when it came to jointly generating energy. The first wind turbine was erected there in 1994, generating 450,000–500,000 kWh of electricity annually. This not only provides the village with clean energy but also provides a return on investment since the village is now an energy supplier[3] (Image 5.4).

Image 5.4 Windmills are a highly effective and environmentally friendly means of generating clean energy (*Source* Pixabay, jwvein)

The Body Shop is also a good example of the refuse strategy: make and sell cosmetics, but refuse to use animals as testing materials.

A notable example of the refuse strategy with a long-standing history is the sale of healthy organic food. Health food stores, which originated from the reform movement, emerged as a response to industrialization, urbanization, and materialism.[4] Nowadays reform is called sustainable and there are many organic products for sale in health food stores and supermarkets.

The sharing economy, discussed in Chapter 4, exemplifies the rethinking strategy of the circular economy. By sharing our cars or tools we intensify their use and fewer people need to buy them.

Thrift shops are an excellent example of the reuse strategy in action. These shops allow you to donate items you no longer need, which are then resold at affordable prices. In the pre-digital era, people would post notes on bulletin boards in supermarkets to sell their belongings, and some stores still maintain such boards. This practice was a precursor to platforms like the Dutch Marktplaats. Another noteworthy initiative is the Repair Café, founded by Martine Postma in 2007. These cafes, which can be found in various locations across different countries (approximately 2400 locations in 2022), provide volunteers who offer their skills to repair broken items.

> 'Repair Cafés are free meeting places and they're all about repairing things (together). In the place where a Repair Café is located, you'll find tools and materials to help you make any repairs you need. You'll also find expert volunteers, with repair skills in all kinds of fields. Visitors bring their broken items from home and repair them with the help of a specialist. It's an ongoing learning process'.[5]

More and more companies offer refurbished products. For example, at rebuy.de you can buy refurbished mobile phones, tablets, cameras, lenses and many other electronic devices for less money than new. You usually receive a warranty on your purchase. This makes this strategy a better option for many people than, for example, Facebook Marketplace. There is also the security of a real company, so you don't have to worry about ordering something that never arrives.

There are many more examples of circular opportunities. For example, did you know that:

- There are plenty of biodegradable cleaning products on the market (and have been for a long time in the form of green soap, for example)?
- Hydrogen is being developed as a carrier for sustainably generated energy?

- There are now many alternatives to plastic, such as bamboo fibre and corn? Just search for bamboo and organic disposable crockery. Great for when you have a party and want to throw the bowls away afterwards.
- You can make new paper from old paper?
- you can reuse coffee grounds as soil for growing mushrooms and making soap?
- restaurants are working to collect their food scraps so they can ferment them to generate energy?
- You can use the search engine Ecosia, which plants a tree for you for every 45 searches.

All these strategies ensure that the harmful effects identified in Chapter 2 can be avoided. In other words, our ecological footprint is reduced by these initiatives.

But product innovation based on circular principles is not enough. If you read the examples carefully, you may wonder whether they are good examples of sustainable products, or whether only certain aspects are sustainable. To be a fully sustainable product, three aspects are important:

- the nature of the raw materials used
- the production processes used
- the effect of the product used (Fig. 5.3).

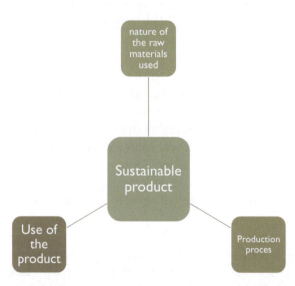

Fig. 5.3 To speak of a sustainable product, three aspects are important

First, sustainable products should either utilize organic materials that can be returned to nature after use or incorporate materials that can be endlessly reused. Second, the production processes must be sustainable. The production of sustainable products should prioritize the use of clean energy sources and adopt environmentally friendly practices. This involves minimizing energy consumption, reducing greenhouse gas emissions, and implementing measures to treat and properly manage wastewater. Additionally, it is important to establish organizations that operate based on principles of equality and social responsibility. Third, sustainable products should have minimal negative impacts when used. This means ensuring that the product's use does not harm the environment or pose risks to human health. For example, even if nitrogen is derived from natural and sustainable sources like cow manure, excessive release can still have detrimental effects on biodiversity. Similarly, using public transportation like trains is generally more sustainable than driving a car, although the sustainability of trains themselves may vary. The same principles can be applied to the batteries used in electric cars. In order for electric vehicles to be fully sustainable, the production and disposal of their batteries should align with the three key aspects of sustainability mentioned earlier. This means using raw materials that are sourced responsibly, minimizing the environmental impact of the battery production processes, and ensuring safe and proper recycling or disposal methods for used batteries. In the next section, you will delve deeper into strategies for making production processes more sustainable. Before that, you will explore two additional ways in which product innovation and improvement can bring benefits to both people and the environment.

A Sustainable Return on Investment

Companies primarily market products that they expect to generate immediate financial gain. This focus on a high ROI hinders the development of sustainable alternatives. Chapter 4 showed the transition we need to make: instead of focusing solely on financial value, it is necessary to broaden our perspective and recognize value in a broader sense. Sustainable companies therefore focus on innovation or improvement that also yield a return for people and the environment. Companies have given this broader view of return various names: *sustainable return on investment* (SROI), *return on environment* (ROE) and *social return on investment* (also SROI). The latter form is sometimes included in tenders by the Dutch government to incentivize selling parties to hire people with disabilities. In this case, a percentage of the total wage amount must be allocated to workplaces for this target group.

India has also established an open source platform for medicines,[6] which aims to develop cures for the most pressing diseases that are available to everyone (Bhardwaj et al., 2011). Since 2011, the *Fix the Patent Laws* campaign[7] has been working structurally to make and keep medicines accessible for people with low incomes.

The Power of Marketing

Marketing is a sophisticated psychological tool employed by companies to persuade customers to make purchases, even for items they may not necessarily need. For example, in recent years, events like Black Friday have been introduced, where companies offer products at purportedly significant discounts. Sustainable companies can also leverage these psychological techniques to attract customers and promote the sale of their products. With marketing, companies can enhance the perception of sustainability and elevate its importance above factors like innovation, quality, and price. Social media is a powerful marketing channel to reach customers. Influencers such as Izzy_manuel (12.6K followers in 2022) do not promote fast fashion, but sustainable garments.

Marketing can play a transformative role by reshaping societal perceptions and challenging prevailing notions of importance. For instance, it can question the glorification of individuals solely based on their wealth and ability to purchase expensive items like luxury watches. Marketing strategies have the potential to expose the true cost behind accumulating wealth, shedding light on the environmental impact caused by excessive business travel and the associated carbon emissions. By highlighting the unsustainability of such behaviors, marketing can help shift societal values towards more responsible and conscientious consumption.

Product innovation and improvement, quality and customer relationship are strategies for companies to work towards a sustainable world. You will learn more about sustainable cost savings in the next section.

5.4 Cost Savings: Making Production Processes More Sustainable

The second strategy employed by companies to remain competitive or maximize profits is cost-cutting. As discussed in Chapter 1, various cost-saving methods were explored, and in Chapter 2, the negative consequences of such practices were examined. Once again, the circular approach presents a solution, as it encompasses not only sustainable product development but also sustainable production processes. Furthermore, in this section, we will delve into the social dimension of sustainable business practices. However, before delving into that, let us examine the correlation between damage and costs, and explore how these costs can be measured.

What Are Costs in the Sustainable Economy?

In today's free market economy, companies incur various costs, which are reflected in their annual accounts. These costs include the purchase prices of raw materials and semi-finished products, personnel costs, material costs such as printing and transportation, as well as depreciation of buildings and machinery. However, in a sustainable economy, the concept of costs extends beyond these financial considerations. In this context, the damage caused by production activities is also considered a cost. For instance, if a company emits CO_2, it incurs a cost associated with the environmental impact. Similarly, if a company purchases materials produced by child labor in India, disregarding the well-being and rights of these children, it incurs a social cost. These costs and damages align in a sustainable economy, emphasizing the need for companies to account for both financial and non-financial impacts.

Companies strive to minimize or avoid damage whenever possible in a sustainable economy. However, in cases where completely avoiding damage is not yet feasible, companies take responsibility by compensating for the costs associated with that damage. A simple example is CO_2 reduction. First, sustainable companies are looking for ways to reduce their CO_2 emissions. Purchasing solar panels is a good way to achieve this. The costs incurred to reduce or avoid damage are called *eco-costs*. A reduction of 1000 kg of CO_2 emissions requires an investment of 116 euros in, for example, wind turbines and solar panels. In other words: the eco-costs of 1000 kg CO_2 are 116 euros.[8] If no further reduction is possible, CO_2 offsetting is an option. The company sets up records of CO_2 emissions and can then offset them annually via Trees for All, for example.

> **Example**
>
> Many organizations let you donate a tree. Examples are The National Forest in the United Kingdom, Trees for All in The Netherlands and Trees for the Future (TREES), active in Africa (Image 5.5).

Image 5.5 Donating trees to offset your carbon footprint (*Source* own image)

The costs involved in this compensation are incorporated into the price. As a result, it is not society as a whole that pays for CO_2 emissions, but the company either settles for lower profits or raises its price, so that only customers pay for the negative externalities. With a higher price, the consumer therefore automatically contributes to planting trees.

There are two ways to make costs and benefits transparent for society:

1. Monetary cost–benefit analysis: costs and benefits are monetized and included in the annual accounts or in a separate statement.
2. Non-monetary cost–benefit analysis: costs and benefits are measured in other units, such as percentages, and are presented in a separate report.

Monetary cost–benefit analysis

The most far-reaching way of including harmful effects in business operations is *full cost* or *true cost accounting*. With this method, selected costs such as CO_2, waste and working conditions are included in the annual accounts and possibly in the price. Full/true cost accounting leads to a true price, which you learned about in Chapter 4. Whether those extra costs are passed on in the market price depends on whether the customer is willing to pay that higher price and on the profit margin on the product. If customers still choose the lowest price, choosing less profit might be required.

> Starting in 2010, PUMA, the manufacturer of sportswear and footwear among other things, developed an example of ecological annual accounts based on the full cost accounting method. It was not easy: PUMA states that it outsources a lot of production, which meant that it also had to investigate the damage caused by the production process with its suppliers. PUMA defined five areas for which the research would take place:
>
> - emissions of CO_2 and other harmful gases affecting the climate
> - water consumption
> - land conversion: conversion of original nature (forest, heath, etc.) into land used for production
> - other air pollutants such as sulphur dioxide, ammonia, nitrogen oxide, carbon monoxide and volatile organic compounds
> - waste.
>
> Eventually, all the detective work led to the presentation of the 2010 annual figures: without all the damage, the company had a net profit of 202 million euros.[9] Including the negative externalities in the annual figures led to a €145 million drop in profit.
>
> | | |
> |---|---:|
> | Net profit | 202 |
> | Damage | −145 |
> | **Sustainable profit** | **57** |
>
> In 2018, PUMA calculated a negative impact of €511 million, 3.5 times higher. Either the figures have become more accurate and comprehensive, or all actions to reduce negative externalities are having little effect.

Two different groups, Global Reporting Initiative and International Integrated Reporting Council, have taken the initiative to support sustainability reporting with integrated *reporting* models. You can compare these initiatives on a macro level to the OECD dashboard, which you learned about in Chapter 4. The models provide themes and guidelines for reporting on sustainability. On 5 January 2023 the Corporate Sustainability Reporting Directive (CSRD) came into operation. It requires companies with certain characteristics to report on sustainability issues, like the impact on the environment and labour practices.

A variant of integrated reporting is the *social cost–benefit analysis* (SCBA). This method has been used for some time in spatial projects and area development by the government, but is also increasingly being applied in health care. The SCBA provides insight into the costs incurred, but above all into the benefits to society. A good example are the activities of social organizations. They employ people who sometimes have difficulty finding work elsewhere. These people often need extra coaching, and sometimes they cannot work at the pace demanded by the market. The social organization therefore incurs extra costs compared with a company that only employs people who can work at the pace required by the market without supervision. At the same time, however, a social organization saves society certain costs: for example, people who work for this company no longer need to receive benefits or debt assistance. Other stakeholders can also benefit. For example, an employee may need to see the doctor less because the work helps them to feel better. A social organization therefore delivers more than just the turnover they generate.

Full/true cost accounting and the SCBA are methods for making sustainable impact financially transparent. These financial methods show that cutting back on one item can have a detrimental effect on another, or vice versa, that certain costs have much more revenue than is visible in the annual accounts. The financial methods therefore provide a more complete picture of all the costs and revenues associated with production.

The disadvantage of these methods is that they are risky. Remember the example of Ford Pinto in Chapter 3? How a human life was monetized, resulting in the wrong decision not to repair the cars? What is value of the life of one bird or of a rainforest that disappears. What value has the growing self-confidence of people who can work for a social organization? Some things cannot be expressed in money, but are part of philosopher Kant's universal rights and duties. However, these methods can still raise awareness about intrinsic value.

Another disadvantage of the methods is that it takes a lot of time to collect all the data and there are few standards for each of the costs. For example, there is a standard amount for CO_2 compensation. This makes these amounts, included in the annual accounts, comparable. But can there ever be standards for a human life?

Non-monetary cost–benefit analysis

In addition to these financial models, a company can also opt for a sustainability report. These reports are less focussed on the financial expression of impact. In sustainability reports you find more objectives in percentages (for example: 'we use 100% recyclable plastic by 2025') and the extent to which these objectives have been achieved.

Companies working on sustainable cost savings often go through the following steps:

1. Identifying themes: what damage does our products and production process create? There are many common negative externalities that every company causes, such as CO_2 emissions and overexploitation. A frequently used tool for measuring environmental impact is the Life Cycle Analysis (see Fig. 5.4). This tool defines the phases from production to sales. Each phase has potential externalities. Questions that can be asked using this tool include: Where do our raw materials come from? Which company extracts them? Are there any data about this company? If not, how can we collect this information?'

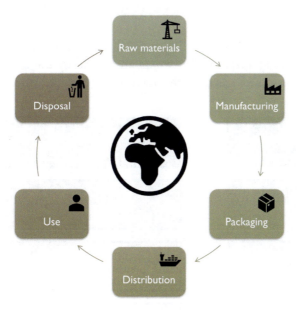

Fig. 5.4 Life Cycle Analysis

2. Draw up a materiality matrix: Which themes are given top priority and which ones must be put on the back burner?
3. Determine unit of measurement: How can we measure a reduction in the theme? For example, if it concerns the production of clothing, a company could use the number of children still working in the factories where it buys its fabrics as a unit. This is where reality comes in: how are you going to measure how many children actually work in the factories and how do you get access to do that measurement? Even if you do get access, what if the factory owner invites you on a day when they leave all the children at home? Measuring harm is a challenging and sometimes impossible job.

4. Zero measurement: In this phase the company investigates how much damage there is at the moment.
5. Setting targets on the themes: Here the company thinks about its goals for the future. For example, in how many years does it want to buy fabrics to which zero children have contributed?
6. Devise and implement ways to meet these targets: How can you ensure that manufacturers reduce the number of children working there? What actions can be devised to reduce CO_2 emissions? This phase is about creativity, but actions can also arise from sound scientific research.
7. Back to phase 4: baseline measurement. Companies examine whether their actions are effective and will achieve the objectives.

For the theme of working conditions in garment factories, this could look as follows (Fig. 5.5).

In this way, a company applies targeted cost reduction to costs that lie outside its own organization, but which are caused by the company. So that ultimately the product reaches consumers with a fair and true price.

Fig. 5.5 Sustainable cost savings

Multiple Cost Reduction: Sustainable Use of Resources

In Chapter 2 you learned that pressure on prices causes producers to try to save on the costs of extracting, processing, transporting and disposing of resources to retain as much profit as possible. Such savings, and with them increased sustainability, can be achieved by using resources more sparingly, intelligently or innovatively. The circularity strategies outlined in section "From Linear to Circular" can help achieve. In this case, efficiency and sustainability coincide.

The Cradle-to-Cradle Institute concretizes this with three possibilities to avoid or reduce damage caused by raw material use:

- The reuse of materials ensures that we need to extract as little new raw material as possible.
- Renewable energy sources can be used during the production process.
- Water is an essential component for all life on earth. By managing water well, a company contributes to a sustainable planet.

The use of raw materials is reduced (more efficient use) by taking back our own products after use. This requires a well-considered product design. In Chapter 4, for example, in the 'from ownership to use' transition, you read very briefly about a bulb manufacturer that no longer sells bulbs, but hours of light. After the bulbs are finished, the manufacturer takes them back to reuse parts. Some carpet suppliers also apply this method, and mobile phone manufacturers are also working on circularity. Additionally, Mudjeans uses your old jeans to make new ones. A final example is the collaboration between IKEA and RENEWI (a Dutch waste management company), which ensures that material from old mattresses is reused in new mattresses.

The consumption of raw materials is also being reduced by the development of the 3D printer. Complete houses can already be built with this innovation. The sustainable claim is that this leads to less waste of raw materials because a 3D printer can determine exactly how much material is needed.

Fossil energy consumption can be reduced by switching to *clean energy*. Earlier you learned about the use of solar panels and windmills. Storage of this energy in hydrogen will ensure that business processes can eventually run on fully sustainable energy. The beauty of this innovation is that it reduces costs and CO_2.

Other high energy costs are often caused by transportation. The logistics sector focusses on optimizing loading to *save transport movements (and therefore CO_2 emissions)*. Step 1 is gaining insight into the loading process. This can be achieved, for example, by sharing loading data with each other.

The use of *clean transport* also naturally reduces CO_2 emissions and improves air quality. CO_2 emissions from transport are reduced by producing as much as possible *close to the customer* (more on this in the remainder of this chapter). The Internet allows conversations with people via Skype or other software, which *reduces the need to travel*. This saves on direct transport costs and lowers CO_2 emissions.

Good water management, so that rivers and seas do not become polluted and drinking water is available for everyone, is the third category mentioned by the Cradle-to-Cradle Institute. Usually governments take responsibility for water purification. But companies can also set up their own installations, so that they can return purified water to nature.

Circularity strategies can also be devised for the use of pesticides, scale, resource cheating and smart tax structures. The use of pesticides (costing the earth, insects and ultimately humans) in food production is increasingly being reduced by switching to *organic farming* and *nature management*. This is not new, but rather a return to how people used to farm: using natural manure instead of artificial fertilizer, making sure the soil is good by growing a variety of products on it (crop rotation) and planting lots of wild flowers on the edges of the fields for optimal pollination of the crops. If there are still fungi and animal pests such as snails, then there are ecological pesticides available which do not harm the crops.

Livestock farming is also classified as organic agriculture. In Chapter 1 you read that food companies try to produce more cheaply through economies of scale: many chickens in one small space, pigs tied up between fences. Large pieces of farmland produce foods that are mainly exported. Sustainable producers and consumers want to turn the tide, which is why many organic producers are switching to *local* and *small-scale production*. The animals are once again given room to roam, and so there are fewer animals on the same piece of land. By producing for the consumer close by (organizing short chains), producer and consumer get to know and trust each other. The consumer's purchases support local employment and reduce the number of kilometres an animal has to travel before it's eaten. Attention is also paid to the feed given to the animals. This is a much healthier environment for the animals, so antibiotics are no longer necessary.

Then you learned about how companies sometimes cheat with the ingredients in their products. Remember the stories about powdered milk and horsemeat? *Providing honest products* and thus stopping cheating require a sustainable mindset from the manufacturer. From the three ethical perspectives, a manufacturer can find reasons to show itself as trustworthy:

- From utilitarianism, a manufacturer stops cheating because it ultimately benefits them more: with a trustworthy reputation, they run less risk of their customers switching to a competitor.
- From duty, the manufacturer realizes that cheating undermines universal duties. If everyone cheated, would that lead to a better world?
- From virtue, the manufacturer decides they want to be a good person and stops cheating because it no longer fits their values.

From the three ethical perspectives, sustainable companies also reconsider their decision to use smart tax structures. Sustainable companies pay *fair taxes* so that their profits contribute to the further development of society as a whole and developing countries in particular.

Equivalence: Joint Decision-Making

In addition to cost savings on the extraction and use of raw materials and other resources, companies also strive for the lowest price by keeping labour costs as low as possible. In Chapter 1, you learned about some of the strategies companies employ to achieve this. In chapter 4 you discovered the importance of sustainable guiding principles to transform harm into positive outcomes, with *equality* at its core. You learned that equality in organizations translates to an equal position in decision-making. Sociocracy was presented as one approach, but democratic decision-making is also effective. Below, you will find a detailed explanation of both variants.

Sociocratic decision-making

Organizations often have various organizational units, such as departments and teams, which are typically managed by department heads or team leaders. However, in sociocracy, things are done differently. Each independent team is referred to as a "circle" in sociocracy. For example, a group of home care workers providing care in a specific neighborhood, a school team responsible for education, or a production team handling the manufacturing of toilet rolls. They make decisions collectively regarding their work.In the sociocratic approach, when a new request for assistance arises in the neighborhood, the circle members decide among themselves who will attend to the person in need and whether adjustments need to be made to an employee's schedule to accommodate it. However, an organization is not solely comprised of production teams. Many organizations also have support staff, such as an ICT department and a finance department, which also form their own circles. What happens when a decision needs to be made that affects both the production circle and the ICT circle? For instance, when the care team requires an app to track the duration of their visits to clients, which also has financial implications. In such cases, decisions are made in the joint circle comprising the chairperson and delegates from both the care and financial circles, as well as the ICT circle if necessary. The basic design of a sociocratic organization is shown in Fig. 5.6.

Hence, the structure is not based on a hierarchical "rake" or pyramid-like hierarchy. Instead, the circles are interconnected through the central circle. Each circle has its own autonomy and decision-making authority within its domain, while also participating in the broader decision-making process through representatives in the central circle. This interconnectedness ensures that decisions are made collectively and that information and feedback flow smoothly between different circles. It promotes a more inclusive and participatory approach to organizational governance. This ground design can be extended according to the different activities.

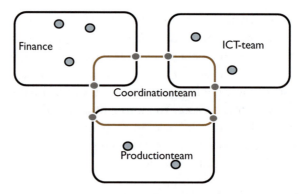

Fig. 5.6 The basic design of a sociocratic organization

> If you want to understand how sociocracy works in practice, the websites of organizations that have implemented it are very enlightening. For example, at the school *De vrije ruimte* in The Hague, the organizational structure is as follows:
> - A separate circle was made for each group of people: one for students, one for parents, one for tutors and one for support.
> - They meet in the school circle through a chair and a delegate.[10]

An important part of sociocracy is the way decisions are made: on the basis of *consent*. Consent means that no one has a serious objection to a proposal. It does not require unanimous agreement or personal preference for the proposal. The key criterion is that no one sees any objection that would prevent the proposal from moving forward. If an objection is raised, it becomes the responsibility of the entire group to address it. Can the proposal be modified to address the objection? Alternatively, can a group discussion lead to a new perspective that prompts the person raising the objection to withdraw it? In sociocracy, any team member can contribute a proposal for decision-making, although often there are designated "owners" who are responsible for specific aspects of the team's work. (for example, for the roster of home care workers or the accreditation of a course).

Decision-making on the basis of consent differs from hierarchical decision-making, where the boss makes a decision for the entire team and the team has to go along with it. Decision-making on the basis of consent is organized according to a clear step-by-step plan in which everyone is given room to say what needs to be said, but must also be quiet if the step-by-step plan demands it. The chair monitors the proper completion of the step-by-step plan.

> **Tip**
> Decision-making based on consent is not easy and requires a learning attitude from all team members. For example, discussing the objections can be complicated: what exactly is the objection and can you listen to it without prejudice (or do you think 'oh, here comes John again with his critical attitude')? What is the proposal to remove this objection and does anyone object to that? If you want to get a feel for the process, you can find some good video's on the internet that give you an idea on how such a process goes.

To make sociocracy work, Sociocracy 3.0 defined seven principles as the basis for the organization[11]:

1. Responsibility: react when needed, do what you agreed to do and take responsibility for the quality of the organization.
2. Equality: involve people in making and developing proposals that affect them.
3. Transparency: make all information accessible to everyone in an organization, unless there is a reason for confidentiality.
4. Effectiveness: only spend time on what brings you closer to achieving your goals (do the rest outside the meeting).

5. Consent: raise, sort out and resolve objections to decisions and actions.
6. Empiricism: test all assumptions through experimentation.
7. Continuous improvement: incremental change to enable learning.

In summary, an organization that operates based on sociocratic principles necessitates a different mindset and behavior from individuals compared to organizations where top-down decision-making prevails. Ownership, responsibility, good communication skills and personal reflection are essential for sociocracy to succeed. In return, the autonomy of each person is maintained, granting employees control over themselves and the issues that pertain to them.

Decision-making based on democratic principles and the principle of consent

The Dutch Breman Installation group has also adopted a decision-making approach based on consent, but they do not utilize decision-making circles and instead employ a democratic decision-making method. The basis for their decision-making method was devised in 1971/2 by one of the founders, Reind Breman. Breman recognized the inherent power imbalance created by capital's control over the workforce and set out to challenge this oppressive system. With the advent of the Breman System, three fundamental principles were established:

– Shareholders and employees have equal say.
– The company is a partnership in which all are basically equal.
– Full responsibility for the company's business is in the hands of the employees.

These principles resulted in a democratic decision-making structure,[12] which has been recalibrated in 2020/2021 to further increase the involvement of all employees. The decision-making system has several components, both at corporate and operating company level.

At the organizational level, decision-making can take two different approaches: the highway for detailed plans and the expeditions that focus on plan development, renewal, and innovation. These approaches are supported by a specialized app called the 'Plek-app'.

In the Expeditions, employees and shareholders are presented with real-world challenges, such as addressing high staff turnover or enhancing work-life balance. They are encouraged to propose their own ideas and solutions.

When it comes to making decisions, the Highway pathway comes into play. Within this route, employees are informed about policy plans that require a decision. Both pathways operate under the principle of consent, meaning that if there are no significant objections, the plan is approved.

However, if there are substantial concerns raised regarding a plan or its progress, employees or shareholders have the right to communicate these issues with management. If their concerns are not adequately addressed, they can request a vote, with a minimum requirement of 15% support. The outcome of the vote will determine the fate of the plan in question. The voting process serves as a safeguard, used sparingly as an exception rather than a common practice.

Individual employees or shareholders were thus indirectly involved in decision-making through their representatives. In 1972, this way of working was satisfactory because the organization was much smaller and the lines of communication were shorter. However, as the organization grew, individual employees and shareholders became less and less involved. The development of the internet, and with it the Plek app, provides an opportunity to involve this group better. The app can reach 1800 employees and 60 shareholders at once. This puts the full responsibility for the company's activities in the hands of the employees.

Within the operating company, staff members are represented by the Works Council (more than 50 employees) or employee representatives (PVT). This representation typically comprises a group of three to seven individuals. However, it is important to note that they do not possess decision-making authority within the organization. The ultimate decision-making power rests with the director of the operating company. The role of the representatives is to provide input, share perspectives, and contribute to the decision-making process, but the final decisions lie in the hands of the director. With regard to consent or advice, the Works Councils Act is largely followed. The Works Council holds the authority to make decisions regarding the appointment and dismissal of the director. This grants the staff members a significant leverage to control the director's power. If the director consistently makes decisions without consulting or obtaining approval from the Works Council/PVT, it may eventually result in dissatisfaction and a decline in support for their position within the organization.

Through these structures, Breman ensured that the organization has a permanently shared decision-making process. No single person, outsider or organizational entity can make decisions individually. Decisions always happen in consultation with all stakeholders in the organization.

Breman is not the only such organization, but few others have implemented shared decision-making at the core of their organization. Another example is the Spanish cooperative Mondragon. According to the definition of the International Cooperative Alliance (ICA), cooperatives are democratically organized,[13] alongside a number of other interesting guiding principles, such as autonomy, independence and economic participation of members.

Does this organizational structure help companies avoid the negative externalities associated with an unequal decision-making system? Can this type of organization contribute to a sustainable economy? Let's reflect on the external costs of the capitalist economic model mentioned at the end of Chapter 2. In an egalitarian organization, all the strategies mentioned can still be employed, but *the significant difference lies in the fact that every employee has the autonomy to decide whether or not to utilize these methods.* A few examples:

- Offshoring: employees are unlikely to choose this because they will lose their jobs.
- Employees have the authority to determine the job structure, including the arrangement of job profiles and associated compensation. In hierarchical organizations, there can be a significant disparity in remuneration between top-level positions and other employees. However, at Mondragon, it has been decided that certain roles within the organization can receive compensation up to six times higher than the lowest-paid position.
- There are also agreements about contracts at Mondragon: after a trial period of one year, the cooperative decides whether someone can become co-owner of the organization (through a one-time payment of, in 2016, 15,000 euros). This gives that person a job for life, as they can no longer be fired.
- If one of the cooperative units of Mondragon is not doing well, the employees will look for positions in another part of the organization.
- If this does not work out either, there is a provision in place where they can remain unemployed for a maximum of two years while receiving 80% of their salary. After the two-year period, they have the opportunity to return to the organization, and if necessary, another individual may take a two-year period of leave.
- Losses are covered by a fund that is established during periods of economic prosperity. This fund serves as a safeguard against cyclical layoffs and rehiring, as well as the need to rely on taxpayer funds during times of recession.

These examples demonstrate the benefits of agreements that are favorable for both the company and its employees. Through mutual consultation, the employees actively participated in the formulation of these agreements, voted on them democratically, and implemented them in practice. While equality is an essential component of addressing harmful effects caused by the unsustainable use of people, it alone may not solve all social problems, such as child labor. It is, therefore, crucial for organizations to work from a comprehensive set of sustainable guiding principles that include ethical awareness and adherence to international standards, such as the International Convention on the Rights of the Child.

Organizing equality thus solves part of the problems. However, we need all the guiding principles of the sustainable economy to operate in a fully sustainable way.

Equality: Everyone Contributes

In a traditional labor market, individuals who cannot keep up with the pace are often marginalized and referred to as "people with a distance to the labor market," emphasizing the labor market's dominance. However, in a sustainable economy, the focus shifts to putting the person at the center rather than the company, leading to the creation of jobs specifically tailored for these individuals. Greyston Bakery is a notable example of a company that embodies this reversal in mission. Their motto, *we don't hire people to bake brownies, we bake brownies to hire people*, highlights their commitment to providing employment opportunities and empowering individuals who may have otherwise been excluded from the labor market.

> The special thing about Greyston is that they hire anyone who wants to work. They call this *open hiring*. Without asking about the past, diplomas, possible drug use. It is not the past that counts, but the future, they say.
>
> Open Hiring fills jobs without judging applicants or asking any questions. It creates opportunities for those who have been kept out of the workforce. That includes women, men, people of colour, people of all faiths and sexual orientations, immigrants and refugees, those living in poverty or who have spent time in prison, and anyone else who has faced barriers to employment. When organizations practice Open Hiring, they upend traditional hiring practices by shifting spending from screening people out to investment in support of employee success.[14]

Greyston Bakery serves as an exemplary social enterprise, demonstrating a business model that prioritizes providing employment opportunities for individuals who may struggle to keep up with the demands of conventional workplaces. Van Orden et al. (2019) argue that the number of social enterprises is growing rapidly. Social enterprises are a great example of the first guiding principle from Chapter 4: people who do fit into the current market system take responsibility for those individuals who, for various reasons, are excluded or marginalized by the system. This behaviour leads to an *inclusive society* in which everyone can participate.

> **Framework of Definitions**
> There is still some ambiguity surrounding the definition of social enterprise. For instance, the website of Social Enterprise presents a broad definition that aligns with the concept of sustainable enterprise. They argue that the development of circular business models, which contribute to a cleaner and more livable planet, also falls under the umbrella of social enterprise since it ultimately affects people's well-being. However, in this textbook, we differentiate between sustainable organizations, which prioritize both ecological and social sustainability, and social enterprises that specifically focus on creating employment opportunities for individuals who face difficulties in finding work.

5.5 Summary

Based on sustainable guiding principles, we have the opportunity to address and prevent the damage caused by our current free market economy. Companies continue to prioritize product innovation, quality, customer relationships, and cost efficiency. However, the key difference is that these objectives are pursued without compromising the well-being of the planet and society.

Characteristics of sustainable business models

1. Companies have a sustainable mission and vision that stems from personal motivation. They have moved from living from the 'what and how' to living from the 'why'.
2. Circular and value-driven product innovation: with product innovation and quality improvement, these companies are making their products circular. They focus on a multiple return on investment.

3. Multiple cost reduction: sustainable companies not only strive to minimize their own costs, but also mitigate the damages caused by their production processes.

Ad 1. Sustainable mission: the 'why' is central

A mission or vision is considered sustainable when its purpose is to create value for both people and the environment. In other words: sustainable entrepreneurs want to lead a meaningful life. They therefore focus on the 'why', and only then talk about the 'how' and the 'what'.

Ad 2. Sustainable product innovation

Sustainable companies invest in making their products circular, focus on a sustainable return on investment and and utilize marketing strategies to promote the popularity of their sustainable products. There are nine strategies companies can follow to make their products more sustainable: refuse, rethink, reduce, reuse, repair, refurbish, remanufacture, repurpose, recycle and recover energy.

Ad 3. Multiple cost reduction

In a sustainable market economy, the damage or negative externalities that companies generate through their products and production processes are recognized as costs.

- To address, minimize, or offset the damage, companies are striving to quantify these costs through various methods. One approach is the incorporation of these costs into financial statements, known as *full cost or true cost accounting*. In this process, all forms of damage are assigned a monetary value (monetization) to enable their inclusion in the annual financial reports.Alternatively, some companies adopt non-monetary objectives and measurements, such as percentages, which are then reported in sustainability reports.

An advantage of making the negative externalities financially transparent is that it provides a more complete picture of all the costs and revenues associated with production. The disadvantage is that these methods are risky and not all things of value can be expressed in monetary terms. In addition, it takes a lot of time to collect all the data and there are few standards for each of the costs.

Sustainable companies minimize damage in the following ways:

1. A sustainable use of resources through circular business strategies (the r-list, see ad 2), including:
   a. recycling of materials
   b. use of renewable energy sources
   c. sustainable water management
   d. organic farming and nature conservation
   e. small-scale and local production (short chain)
   f. fair products
   g. paying fair taxes in the country where they produce.

2. Equality: joint decision-making. In an equal setting, the company is run collectively, emphasizing the importance of shared decision-making. You have explored two specific organizational approaches:

   - sociocracy
   - democratic organizing.

   Because everyone participates, decisions are made that are good for everyone, rather than just for the top management and shareholders.

3. Equality: everyone contributes. Instead of prioritizing the product, the emphasis is on the individual, designing a product that aligns with their capabilities. Enterprises that operate with this vision are classified as social enterprises, where equality and inclusivity are fundamental principles.

Notes

1. Council of State: Dutch approach to nitrogen is flawed. Accessed October 2, 2022, from https://nos.nl/artikel/2286818-raad-van-state-nederlandse-aanpak-stikstof-deugt-niet.html.
2. Tegelicht: Waste is food. Accessed October 2, 2022, from www.vpro.nl/programmas/tegenlicht/kijk/afleveringen/2006-2007/afval-is-voedsel-deel-1.html.
3. Village mill Reduzum. Accessed October 2, 2022, from www.dorpsmolen-reduzum.nl/over-ons.
4. Reform Movement. Accessed October 2, 2022, from https://en.wikipedia.org/wiki/Lebensreform.
5. Repair Cafe. Accessed September 26, 2022, from https://www.repaircafe.org/en/about/.

6. Open Source Drug Discovery (OSDD). Accessed October 2, 2022, from www.OSDD.net.
7. Fix the patent laws. Accessed October 2, 2022, from www.fixthepatentlaws.org/.
8. If you want to learn more about eco-costs, the TU Delft website is a good starting point: www.ecocostsvalue.com/ (accessed 2 October 2022).
9. See here the annual figures of PUMA: https://about.puma.com/en/sustainability/reporting. Because in the 2010 report the 145 million Euros cannot be found properly, we also used this report, which unfortunately can now no longer be found on PUMA's own site: http://danielsotelsek.com/wp-content/uploads/2013/10/Puma-EPL1.pdf (accessed October 2, 2022).
10. Sociocracy: Every vote counts. Accessed October 2, 2022, from www.devrijeruimte.org/ons-onderwijsdemocratisch-onderwijs/sociocratie/.
11. The seven principles. Accessed October 2, 2022, from https://sociocracy30.org/the-details/principles/.
12. Their previous method was previously published in the *Monthly Journal of Accountancy and Business Economics*: From top-down management control to democratic decision-making. Accessed October 2, 2022, from https://research.hanze.nl/en/publications/van-top-down-management-control-naar-democratische-besluitvorming.
13. Cooperative identity, values & principles. Accessed October 2, 2022, from www.ica.coop/en/cooperatives/cooperative-identity.
14. Greyston. Accessed October 2, 2022, from www.greyston.org.

Literature

Bhardwaj, A., et al. (2011). Open source drug discovery—A new paradigm of collaborative research in tuberculosis drug development. *Tuberculosis, 91*(5), 479–486.

van Orden, C., Irosemitho, M., & Teesink, F. (2019, June). What do you portray as social impact? *Social Bestek, 81*, 38–41.

Afterword: Taking Action! The Role of Individuals, Businesses and Government

In this book you have learned what market forces means, and why this fundamental system in our economy leads to unwanted effects.

A sustainable free market can experience growth when three key stakeholders take action: individuals, businesses, and government. If you value sustainability, you have the power to make a difference in all three areas.

1. As an individual, you make daily choices about where you spend your money. With this money you determine what is produced: if many people continue to buy non-sustainable products, non-sustainable organizations will continue to exist. Consumers can make or break a company, as you could read in Chapter 1 in the case study about Nokia's mobile phone. In 1989, Dutch comedian Youp van 't Hek delivered a year-end conference where he openly criticized non-alcoholic beer, specifically targeting Buckler. As a consequence of his remarks, Buckler sales experienced a significant decline as consumers didn't want to be associated with the negative perception attached to the brand. However, times have changed, and non-alcoholic beer has evolved to become an integral part of the market. So inform yourself about the origins of what you buy, find sustainable alternatives and change the world with other consumers.
2. Many of you will find yourselves working within an organization or perhaps even starting your own company. In these roles, you have the opportunity to become a leader in driving sustainable strategies. Take the time to examine the mission and vision of your organization: are they

oriented towards enhancing the well-being of both people and the environment? A sustainable organization embraces circular business processes, where resources are utilized efficiently and waste is minimized. Furthermore, within such an organization, individuals are treated as equals and share responsibility for matters that affect them. Chapter 5 presents more components of a sustainable business model. If you want to make an organization sustainable, you can take the following six steps:

a. Convince management of the value and necessity of implementing a sustainable strategy and operations (if equal relationships are already established, you can skip this step and proceed to step d).
b. Form a dedicated group with your colleagues who are committed to driving sustainability within the organization.
c. Collaborate with this group to develop a sustainable agenda for the next five years, setting specific objectives and allocating a project budget. Seek inspiration from other organizations, engage in dialogues with them, and study their sustainability reports. Remember, you don't have to reinvent the wheel. Foster support within the organization by encouraging colleagues to share their ideas and determine feasibility. Consider organizing sustainability meetings and establishing a social media account for colleagues to contribute their ideas.
d. Present the plan to the board and secure their commitment to its implementation.
e. Appoint leaders for each sub-project and begin working towards achieving the established objectives.
f. Create a powerful marketing campaign towards the customer and other stakeholders. Can you think of a Buckler variant that will make the customer switch en masse from the non-sustainable product to the sustainable alternatives offered by your organization?

3. The third way to create impact is through politics. With your vote, you have the power to shape government policies. Take the time to study the party platforms and identify which party or parties are committed to building a sustainable nation. Remember that sustainability requires a harmonious balance between the well-being of the planet, people, and profitability. Be discerning of parties that make promises of a sustainable paradise without considering the financial implications. Also, be cautious of parties that scapegoat certain groups of people for societal issues, as market forces often play a significant role in shaping these challenges.

What impact do you want to have with your life?

Literature

Achterhuis, H. (2010). *The utopia of the free market*. Lemniscaat.
Adams, R. B., Licht, A. N., & Sagiv, L. (2009). *Shareholderism: Board members' values and the shareholder-stakeholder dilemma*. CELS 2009 4th Annual Conference on Empirical Legal Studies Paper.
Ariely, D., Gneezy, U., Loewenstein, G., & Mazar, N. (2009). Large stakes and big mistakes. *Review of Economic Studies, 76*(2), 451–470.
Bhardwaj, A., et al. (2011). Open source drug discovery—A new paradigm of collaborative research in tuberculosis drug development. *Tuberculosis, 91*(5), 479–486.
Blasi, J., Kruse, D., & Freeman, R. B. (2018). Broad-based employee stock ownership and profit sharing: History, evidence, and policy implications. *Journal of Participation and Employee Ownership, 1*(1), 38–60.
Block, P. (1993). *Stewardship. Choosing service over self-interest*. Berret-Koehler Publishers.
Blom, N. (2019). *Partner relationship quality under pressing work conditions: Longitudinal and cross-national investigations*. Ipskamp Printing.
Bodirsky, B. L., Chen, D. M.-C., Weindl, I., Soergel, B., Beier, F., Molina Bacca, E. J., Gaupp, F., Popp, A., & Lotze-Campen, H. (2022). Integrating degrowth and efficiency perspectives enables an emission-neutral food system by 2100. *Nature Food, 3*(5), 341–348.
Brothers, A. (2013). *Money and good: Lessons for benevolent capitalists*. Skandalon.
Corbey, M. H. (2010). Agent or steward? On human view and management control. *Monthly Journal of Accountancy and Business Economics, 84*(9), 487–492.

Literature

Deci, E. L., Koestner, R., & Ryan, R. M. (1999). A meta-analytic review of experiments examining the effects of extrinsic rewards on intrinsic motivation. *Psychological Bulletin, 125*(6), 627–669, discussion: 692–700.

Deci, E. L., Olafsen, A., & Ryan, R. (2017). Self-determination theory in work organizations: The state of a science. *Annual Review of Organizational Psychology and Organizational Behavior, 4,* 19–43.

Elkington, J. (1999). *Cannibals with forks: The triple bottom line of 21st century business.* Wiley.

Gagné, M., & Deci, E. L. (2005). Self-determination theory and work motivation. *Journal of Organizational Behavior, 26*(4), 331.

Georges, A., & Romme, L. (1999). Domination, self-determination and circular organizing. *Organization Studies, 20*(5), 801–832.

Goudzwaard, B., & de Lange, H. M. (1991). *Enough of too much, enough of too little.* Ten Have.

Hardin, G. (1968). The tragedy of the commons. *Science, 162*(3859), 1243–1248.

Hurst, A. (2017). *The meaning economy.* Scriptum.

Lucassen, J. M. W. G. (2013). *The history of work and labour.* ISHH Paper. https://iisg.amsterdam/files/2018-01/outlines-of-a-history-of-labour_respap51.pdf

Marx, K. (1867). *Capital.* Downloaded as an e-book from Marxist.org

Piketty, T. (2014). *Capital in the twenty-first century.* Belknap Press of Harvard University Press.

Porter, M. (1985). *Competitive advantage: Creating and sustaining superior performance.* Free Press.

Rigby, C., & Ryan, R. (2018). Self-determination theory in human resource development: New directions and practical considerations. *Advances in Developing Human Resources, 20*(2), 133–147.

Singh, S., & Aggarwal, Y. (2018). Happiness at work scale: Construction and psychometric validation of a measure using mixed method approach. *Journal of Happiness Studies: An Interdisciplinary Forum on Subjective Well-Being, 19*(5), 1439–1463.

Smith, A. (1784). *An inquiry into the nature and causes of the wealth of nations.* MetaLibri.

Smith, A. (1790). *A theory of moral sentiments* (6th ed.). MetaLibri.

Speklé, R. F., & Verbeeten, F. H. M. (2014). The use of performance measurement systems in the public sector: Effects on performance. *Management Accounting Research, 25*(2), 131–146.

Treacy, M., & Wiersema, F. (1995). *The discipline of market leaders.* Addison-Wesley.

van Bavel, B. (2018). *The invisible hand.* Promotheus.

van den Broeck, A., et al. (2009). The self-determination theory: Qualitative motivation in the workplace. *Behavior and Organization, 22*(4), 316–334.

van der Stede, W. A. (2011). Management accounting research in the wake of the crisis: Some reflections. *European Accounting Review, 20*(4), 605–623.

van Orden, C., Irosemitho, M., & Teesink, F. (2019, June). What do you portray as social impact? *Social Bestek, 81,* 38–41.

van Witteloostuijn, A. (1999). *The anorexia strategy*. De Arbeiderspers.
Verhaeghe, P. (2013). *Identity*. De Bezige Bij.
Vosselman, E., & de Loo, I. (2018). Performativity in networks: The Janus head of accounting. *Monthly Journal of Accountancy and Business Economics, 92*(1/2), 21–25.
Wawoe, K. W. (2010). *Proactive personality: The advantages and disadvantages of an entrepreneurial disposition in the financial industry*. www.WorldCat.org
Wilkinson, R. G. (2005). *The impact of inequality: How to make sick societies healthier*. Routledge.
Zander, B., & Zander, R. S. (2002). *The art of possibility*. Penguin Books.

Index

A

alienation 48, 52, 61, 77, 120
A theory of moral sentiments 101
autonomous motivation 82, 84
autonomy 77, 82, 124, 164, 166

B

bio-based 145
biodiversity 45, 60, 61, 129, 139
biomimicry 145
Brundtland viii, 32, 100, 109

C

capitalism 78, 83
circular 141, 144, 146, 149, 150, 168–170
circular economy xv, 145, 147
circularity strategy(ies) 145, 159, 160
coercive hand of the free market 12
competence 53, 82, 124
competitive position 8, 12, 14, 15, 24, 25, 29, 92

Conference of the Parties 130
consent 163–165
consumer society 7
controlled motivation 82, 83
COP. *See* Conference of the Parties
cost leadership 5–7, 25
cradle-to-cradle 159, 160
customer intimacy 6, 7, 24, 25

D

differentiation 5–7, 24, 25
downcycling 145
duty ethics 107

E

eco-costs 153
ecological footprint 32, 60, 145, 150
economic growth vii, viii, 14, 15, 32, 56, 68, 71, 84, 85, 88, 89, 95, 100, 124, 126, 127, 133
economic recession 54, 89
efficiency 13, 14, 16–18, 47, 50, 71
empathy 69, 101, 102

182   Index

ethics 99, 103–105, 121, 132, 134
externalities 61, 62, 92, 118, 120, 144, 154, 155, 157, 169

F

fair price 118, 119, 129, 133
fast fashion 152
focus. *See* customer intimacy
free market v, vi, 1, 2, 4, 8, 13, 14, 24, 25, 42, 67–69, 73, 81, 82, 91, 94, 95, 101, 113, 132, 153
free market economy 168
free trade 68, 71, 91, 94, 113, 114

G

GDP. *See* Gross Domestic Product
greenwashing 140
Gross Domestic Product 85, 126
gross national happiness 126

H

hierarchical structure 78, 80
human scale 52, 64

I

impact investing 119
impartial spectator 101
inclusive society 168
invisible hand 4, 29, 69, 91, 94

J

job specialisation 71

L

liberalism 69, 91
linear use 31, 32

M

marketing 3, 16, 17, 25, 26, 29, 33, 41, 152, 174
Marx. *See* Marx, Karl
Marx, Karl 48, 78, 120
means of production viii, 17, 18, 72, 76, 81, 120, 123
mimetic desire 33
monetary value 74, 95, 103
multiple value 118, 120, 138

N

negative externalities 169
neoliberal 92

O

operational excellence. *See* cost leadership
opportunistic 82, 95
overshoot day 32

P

Piketty. *See* Piketty, Thomas
Piketty, Thomas 53, 55, 120
plastic soup 40, 144
private equity firms 10
private property 68, 72, 73, 81
productivity 17–22, 25, 26, 46–48, 50, 53, 68, 70, 71, 82, 94, 112, 121
product leadership 5–7, 13, 14, 24, 25
profit maximization 10–14, 31, 41
public organisation 73, 91
purpose economy 105

R

return on investment (ROI) 32, 41, 148, 168
ROI. *See* return on investment

S

self-determination theory 82
self-interest 21, 22, 30, 31, 69, 72, 81, 95, 100, 102, 111, 132
sharing economy 115, 116, 133
Smith. *See* Smith, Adam
Smith, Adam v, 4, 21, 22, 67, 69, 70, 101, 102, 105, 106
social enterprise 168
social inequality 46, 61, 62, 78, 79, 93, 120
sociocracy 123, 124, 161–164, 170
spectator 101
surplus 57, 76, 79, 89, 95, 121, 130
Sustainable Development Goals 110, 119
sustainable market economy v, vi, xvi, 68, 113, 120, 125, 132, 137, 169

T

The wealth of nations 69, 101
too big to fail 51

Tragedy of the Commons 30, 31, 41, 115
trickle down 55, 68, 84, 88
triple bottom line 107
true price. *See* fair price

U

Universal Declaration of Human Rights 107, 122
upcycling 145
use value 41, 74

V

value for money 3, 15, 16

W

wicked problem 112

Z

zero sum game 69

Printed in the United States
by Baker & Taylor Publisher Services